MAKING TRUTH

MAKING

TRUTH

METAPHOR IN SCIENCE

Theodore L. Brown

University of Illinois Press
Urbana and Chicago

Unless otherwise indicated, the
figures reproduced in this book
were prepared by Janet Sinn-
Hanlon of the Imaging Technology Group, Beckman Institute
for Advanced Science and Technology, University of Illinois at
Urbana-Champaign.

Library of Congress Cataloging-
in-Publication Data
Brown, Theodore L.
Making truth : metaphor in
science / Theodore L. Brown.
p.   cm.
Includes bibliographical references and index.
ISBN 0-252-02810-4 (cloth : alk.
paper)
1.Science—Philosophy.
2. Metaphor. 3. Interdisciplinary
approach to knowledge. I. Title.
Q175.B7966   2003
501—dc21   2002011212

Truth cannot be out there—cannot exist independently of the human mind—because sentences cannot so exist, or be out there. The world is out there, but descriptions of the world are not. Only descriptions of the world can be true or false. The world on its own—unaided by the describing activities of human beings—cannot.

—Richard Rorty,
  Contingency, Irony, and
  Solidarity

CONTENTS

PREFACE

This book has been a long time in the making. In 1985 the University of Illinois at Urbana-Champaign received a $40 million gift from Arnold and Mabel Beckman to establish a large multidisciplinary research institute. I was put in charge of the effort and eventually became the founding director of the Arnold and Mabel Beckman Institute for Advanced Science and Technology.

When it opened its doors, the Beckman Institute was, and probably still is, the largest and most broad-based interdisciplinary research facility in any academic setting. The idea was to bring together researchers from a wide range of disciplines to work on research problems of a broad, multifaceted nature, spanning several disciplines. The Beckman Institute faculty come from departments including physics, chemistry, physiology, cell biology, psychology, linguistics, electrical engineering, computer sciences, and philosophy.

The institute was designed to promote interactions between faculty, undergraduate students, graduate students, postdoctoral researchers, and staff. To break through the traditional barriers to collaborative work across departmental boundaries, we designed the building to be beautiful as well as functional and to promote interactions between people with diverse backgrounds. The institute opened its doors in 1989. Since then, the experiment in a new way of conducting interdisciplinary research in an academic setting has succeeded beyond everyone's ambitious expectations.

But why should it succeed at all? Why should we have imagined that by mixing a diverse lot of researchers in a large, shared facility we would see something happen that is more than the sum of a collection of talented researchers going their own ways? Good intentions alone would not make the institute a success. We learned that putting people together in well-designed surroundings, with institutional encouragements to work across the traditional departmental boundary lines, made for a bit of magic.

So if it's magic, wherein does the magic arise? What ingredients make up the added value of a freely interdisciplinary environment? I soon realized that exchange of significantly different views of the world between researchers from diverse disciplines is an important factor. The different views of a particular problem arise from the prevailing metaphors held by each discipline. Sharing different metaphorical representations of a problem appears to open up possibilities for creative thinking. It was natural enough to focus on the roles of metaphor; among the cognitive scientists who were initial faculty residents in the Beckman Institute were Ken Forbus, Dedre Gentner, Jerry Morgan, and Andrew Ortony. It was Jerry Morgan who put me onto *Metaphors We Live By*, by George Lakoff and Mark Johnson.

I saw early on that before I could answer my original questions I would need to understand the nature of metaphor more generally. The landmark treatises on metaphor, notably those by I. A. Richards and Max Black, dealt with metaphor largely as manifested in figurative language, such as poetry and fiction. However, a proper appreciation of metaphor's roles in science comes from understanding its conceptual nature, and this entails a more comprehensive understanding, grounded in linguistics and psychology. It is just this kind of understanding that has been made possible by the development of cognitive sciences as an interdisciplinary field of study in its own right.

My approach in this book has been to show that metaphor plays a central role in the development of a scientific subject, from its very beginnings through to its full development as a mature body of knowledge and understanding. It figures in the scientist's initial creative impulses, in interpretations of experimental data, in formulations of scientific explanations, and in communication between scientists and between scientists and the rest of the world. To show this, I have chosen a wide range of examples, from the venerable and simple case of the atom to current hot topics of investigation, such as protein folding and global warming. The treatments in all these cases are at a level that demands no prior special knowledge of the science. The emphasis is on showing the roles of metaphor at all stages of development. Before one can discern the metaphorical nature of scientific reasoning and communication in these diverse examples, it is necessary to understand the basics of the theory of conceptual metaphor. One of the early chapters of the book therefore presents a brief account of the theory as a basis on which to follow the developments in subsequent chapters. I find this a fascinating subject. Whether or not you share my enthusiasm, I hope you will bear with me in reading chapter 3; it will help you appreciate the chapters that follow.

It is impossible to deal with the subject of metaphor in science without touching upon matters that occupy the attentions of historians, philosophers, and social scientists interested in science. Although this book is not about the so-called science wars, scientific realism, or constructivist views of science, all these topics are touched upon. I believe that an appreciation of the central role metaphor plays in all aspects of doing science leads one to important conclusions about the status of scientific knowledge, about realism and reductionism, that simply don't fit with the views held by many scientists and others.

* * *

I want to thank Garin Brockman, John Brockman, Audrey Brown, Mary Coffield, Dedre Gentner, Adele Goldberg, Atlee Jackson, Kirk Jensen, Jay Labinger, George Lakoff, Zan Luthey-Schulten, Greg Murphy, Ken Suslick, Bob Switzer, John Walsh, and two anonymous reviewers for helpful advice and constructive criticism. The faults that remain are all mine. Janet Hanlon has done a fine job of producing a majority of the book's figures. I want to thank also the helpful people at the University of Illinois Press, notably Richard Martin, executive editor; Theresa L. Sears, managing editor; Copenhaver Cumpston, art director, and Carol Anne Peschke, copy editor. Finally, I acknowledge support from the Beckman Institute at the University of Illinois.

MAKING TRUTH

My aim in this book is to present a new perspective on the ways in which science gets done. That means a new perspective on how scientists reason about the world, design and interpret experiments, and communicate with one another and with the larger society outside science. From many years of experience as a practicing scientist and science administrator, I have been forced to conclude that most people, including many scientists, don't have a clear understanding of the relationship of scientific knowledge to other forms of knowing. This means that they don't fully appreciate how science works. Among nonscientists, there is a widespread feeling that it doesn't pay to be too curious about such matters, that science is something of a closed community, not readily penetrated. For many, science is viewed as a form of magic that can be practiced only after long apprenticeship. What scientists actually do and how they go about doing it are mysteries. Many scientists share a different set of misconceptions. They take for granted many claims about the rationality and objective character of science that do not hold up well under critical scrutiny.

If science and the products of scientific activity were not so important in modern life, misconceptions about its nature would not matter so much. But science is important, in both intellectual and practical terms, to those in whom it generates a sense of fear and loathing as well as to its champions. To properly appreciate its place in contemporary life, we must understand more fully how science really functions—what is involved at the cognitive level in scientific discovery and the communication of scientific

1

SCIENTIFIC THOUGHT AND PRACTICE

results. When we comprehend the bases on which scientists reason about nature, interpret observations, and communicate with one another and with the rest of the world, we will be in a better position to evaluate science in relation to other fields of intellectual activity.

Consider the great progress made recently in mapping the human genome. These exciting developments in basic science have given rise to a host of public policy questions with important ethical and economic implications. For example, what kinds of genetic testing should be allowed, and under what conditions? How, if at all, should genetic information on an individual citizen be made available to others? Because these are questions of public policy, people should have a voice in addressing them. But to understand the issues and to express their views effectively, people must be able to comprehend relevant scientific information.

For most nonscientists, the science courses taught in schools, from elementary through university, are the means of access to knowledge of science. The way in which science is taught in school has a lot to do with students' understanding not only of the content of science but of how science compares with other forms of knowledge. It helps to shape their comprehension of science as a social and intellectual endeavor within the larger framework of modern society. Just as important as the specific facts and concepts of science are ideas of how and where those facts and concepts arise, the nature of the truth they represent, their authority or contingency, and their relationship to other forms of knowledge.

A recent newspaper article reported that a committee of the American Association for the Advancement of Science, charged with evaluating high school biology texts, gave unsatisfactory ratings to all of the ten most widely used textbooks. The committee judged that the texts emphasized mere memorization and failed to encourage students to examine their ideas or relate the lessons to hands-on experiments or everyday life. Put simply, they tended to miss the big picture. If the text materials used in science education are defective, it is quite likely that the appreciation and understanding of science that is expected to derive from their use will be defective as well.

This book is concerned with an aspect of science that is largely absent from textbooks yet is central to an understanding of the nature of science. I want to show that much of what scientists do—how they conceive of productive experiments, what they observe, and their interpretations of observations—is governed by metaphorical reasoning.

Science in Western Society

Everyone acknowledges that modern life owes much of its character to the twin forces of scientific discovery and technological development. Technology shapes our view of the world because its products change the physical conditions of our lives and the ways in which we interact with the world. By contrast, science continually presents us with new stories, new visions, of the very nature of the world. There is perhaps no better marker for the beginnings of the modern era of science than Galileo Galilei's book *The Sidereal Messenger.* Published in 1610, it recounted Galileo's telescopic observations of the moon and the planets, particularly Jupiter. His accounts of many new phenomena challenged conventional wisdom regarding the nature of the heavens and forced new conceptions of the nature of the world.

Galileo's observation that Jupiter's moons revolve around it contradicted the conventional wisdom that all the heavenly bodies revolve around the earth. Since Galileo's time, new theories and systematic observations of nature have continued to challenge entrenched beliefs about the natural order. Johannes Kepler's laws of planetary motion, based on accumulated observations of the apparent motions of the planets, did away in one sweeping gesture with the traditional foundations of astronomical theory, based on Ptolemy's and Copernicus's idea of circular orbits. Einstein's special theory of relativity spelled the end of the Newtonian concept of time and its relationship to space.

In Western culture, the period dating from the seventeenth century often is marked as the beginning of the Scientific Revolution. Philosophers were forced to reconsider humankind's relationship to nature and to ask how and to what extent we could learn about the external physical world. Today, the ideas, practices, and theories of scientists and the values they hold are potent elements in the affairs of society. The products of scientists' work have become central to modern life. For example, the revolution in molecular biology has already triggered great changes in biomedical practices, and even greater changes are in the offing. Because scientific advances can affect many important aspects of daily life so profoundly, everyone, not just the scientists themselves, has a stake in what scientists discover and in the applications of those discoveries. We really ought to care about how science is taught in our schools. And part of the science curriculum should be concerned with how scientists come to know what they purport to know and the reliability of that knowledge.

How Do Scientists Work?

Nearly everyone would agree that science matters a great deal to human society. Like every other component of modern culture, science therefore comes in for its share of scrutiny. What are scientists studying, and for what reasons? Who is paying for their work? What are the implications of a new discovery or technological development in terms of potential effects on the economy, health, and the environment? What risks and benefits are involved in any particular public policy decision? Should there be limitations on what scientists can study? If the answer is yes, how are those limits to be decided?

Most people are not very sophisticated about science. They lack knowledge of the content of most areas of science, of the accepted paradigms of fields as diverse as cell biology, neurophysiology, natural product chemistry, and particle physics. The amount of information in the sciences is so vast that even most scientists are largely ignorant of fields other than their own specialty. Given the enormous extent of the scientific literature, everyone—scientists included—is fated to depend on someone else to select, summarize, and evaluate what is known about scientific topics that come up for discussion. But preceding the question of how to evaluate specific bodies of scientific observation and theory, there are more general questions of how scientists come to know and understand the world.

In pondering the significance of what scientists have to say, one of the first questions to ask is, What special character, if any, does scientific knowledge have? Clearly, we gain knowledge of the world through the use of our senses and such extensions of our senses as we might devise, such as a telescope or microscope. But what do we see? Is there an objective, mind-independent reality about which we can obtain some measure of knowledge? If so, what is the nature of that knowledge? What limitations apply to what we might learn? Is our capacity to reason and hold ideas of the world in some way distinct from our brain and sensory system? Are we capable of holding ideas that are in some way less fallible than those based on sense data alone? A long line of philosophers, from Descartes and Kant through the analytical philosophers, to their contemporary successors, have considered such perennial questions as whether the physical world exists as an objective, mind-independent entity and whether—and to what extent—we can come to know it. The philosophers have not managed to arrive at anything like unanimity on this and many related questions.

Although the answers have proven difficult to come by, the questions philosophers ask about the nature of our knowledge of the physical world are more than mere fodder for academic debates; they are relevant here and now. Scientists present ideas and prospects for the manipulation of nature in an ever-increasing stream. Many of these have deep and potentially disturbing social, cultural, economic, and environmental implications. For example, many scientists have been warning for some time that within the next fifty years or so, the continued burning of fossil fuels will cause sufficient global warming to substantially modify Earth's climate. Are their predictions correct? How do we know? How do they and we go about evaluating the work that underlies these predictions? As another example, it is argued on one hand that genetically modified plants are a boon to agriculture and a safe way to increase productivity while reducing reliance on pesticides and herbicides. Others claim that these genetically modified organisms pose a threat to human health and the environment. What is the basis on which either side makes its claims? What truths are contained in these positions?

We typically have extensive bodies of evidence bearing on the sorts of scientific questions that surface as public policy issues. But at a deeper level, there are questions of the sort that the philosophers worry about. What can we say about how scientists know what they claim to know? Is there something special about scientific knowledge that sets it apart from other forms of knowing, from other intellectual pursuits? Many scientists are sure that science is special, that there is a characteristic scientific method by which new knowledge is garnered and evaluated and by which hypotheses and theories are formulated and tested. The discipline of scientific methods, they argue, leads to a more objective kind of knowing. The scientist can approach true insights into the nature of a mind-independent reality, the world as it really is. Others hold the quite different view that scientists can gain only a limited and largely subjective understanding of nature.

Any attempt to gain perspective on long-standing questions of the nature of scientific knowledge must take account of the fact that science is, among other things, a social enterprise. We must understand that science is more than individuals working away in splendid isolation. It is an intellectual pursuit, of course, but it is also a social activity, a job, a pathway to some form of social and economic status, and so on. New scientific knowledge is generated in the context of interactions between scientists. The motivations in conducting scientific research are many, and they bear on the ways and extents to which new discoveries are communicated within the scientific community and to the larger society. This means that we must consider all the factors that go into making up the scientific enterprise if we are to understand and evaluate what science produces.

What Constitutes Valid Science?

In this section I want to briefly trace the development of some ideas current in the twentieth century on the nature of scientific thought and practice. Some of the people I have chosen to cite are not necessarily the best-known figures, but their contributions and place in time reveal the evolution of views on scientific reasoning and creativity that are of special significance for us.

By the end of the first half of the twentieth century, the dominant philosophical positions with respect to science fell under the broad title of logical empiricism or logical positivism. The major emphasis in these philosophical schools was on the structure of scientific theories. There was a strong reliance on the techniques of mathematical logic. The idea was to treat scientific theories in roughly the same way as mathematical theorems. In theories there are axiomatic statements—the assumptions of the theory—and statements linking various terms in logical ways. The terms in any theory were required to correspond precisely to observables. Those observables in turn were assumed to correspond precisely to things as they are in the world. Thus, formally at least, scientific theories structured in this way can aim to provide accurate descriptions of an objective world, independent of human categories and influence.

The logical empiricist tradition proved to be a difficult and constraining regimen. Few scientific theories could come close to meeting its formal strictures; perhaps none could actually achieve a state of completeness. Most seriously, because it dealt with formal relationships in much the same way as abstract mathematical formulations, logical empiricism was divorced from actual scientific practice. Part of the rhetoric of logical empiricism was that it was not concerned with the details of how science got done or with the social or psychological factors that formed the milieu in which the scientist worked. As the saying went, "There is no logic of discovery," there is only the logic of justification, that is, the logical construction and evaluation of theories.

The distancing of philosophical thought from scientific practice was not lost on some philosophers of the time. Karl Popper, who has exerted a great deal of influence since the 1930s, took the view, in opposition to the logical empiricists, that a theory could never be proven to be correct in all respects, could never be shown to be logically complete.[1] However, it could, at least in principle, be falsified. That is, one should be able to design an experiment that could yield a result at variance with the predictions of a particular theory. If the result of the experiment is in accord with the theory, the theory is as a result more securely established, although it remains a necessarily contingent explanation. If the result is at variance with the theory, the theory is thereby falsified and must be replaced.

Popper's concerns with scientific methods brought his outlook closer to scientific practice. But he rejected induction as a method for verifying or arguing the truth of any general law. He took for granted the capacity of experiments to provide accurate information about the objective, real world. He had little interest in the contexts in which scientists work or in the ways in which scientific results are applied in the real world.[2] He thus failed in largely the same ways as the logical empiricists to relate in a meaningful way to what scientists actually did and how they did it.

In 1964 John R. Platt published a paper in the journal *Science* titled "Strong Inference," in which he argued for an approach to science that he saw as a variant of Francis Bacon's inductivism:

Strong inference consists of applying the following steps to every problem in science, formally and explicitly and regularly:

1. Devising alternative hypotheses;

2. Devising a crucial experiment (or several of them), with alternative possible outcomes, each of which will, as nearly as possible, exclude one or more of the hypotheses;

3. Carrying out the experiment so as to get a clean result;

1'. Recycling the procedure, making sub-hypotheses or sequential hypotheses to refine the possibilities that remain; and so on.[3]

Platt was not making a claim for originality in this formulation, which he attributed to T. C. Chamberlin, a turn-of-the-century geologist. His article was more of a provocative assertion of how science should be done and a criticism of much scientific practice. Too many scientists, he argued, are intellectually lazy, have become method oriented rather than problem oriented, and are given to simply collecting data rather than designing experiments that decisively resolve important open questions.

Platt's effort to spell out how good science should be carried out contains many of the elements of Popper's philosophy[4] even as he stepped beyond the latter's more formal and constrained formulation. However, his insistence that "strong inference" was the only effective way to do science prompted irate letters to the editor in later issues of *Science*. As had Popper before him, Platt failed to take into account many of the factors that constitute productive scientific practice. For example, he was unable to appreciate the motivation for a program of chemical synthesis or many biological field studies. Although strong inference may be the most efficacious procedure in a particular area of science or with respect to a particular problem, there is much more involved than any single method in producing, validating, and communicating scientific results.

A Turn toward Questions of Practice

Michael Polanyi, a British scientist and philosopher, began in about 1950 to consider the nature of scientific activity.[5] As an experienced scientist of outstanding accomplishment, Polanyi identified the shortcomings of the logical empiricist formulation of science. He also rejected extensions such as those proposed by Popper on the grounds that all such approaches demand an unattainably rigorous correspondence between scientific generalizations and the world. While maintaining a degree of realism, he recognized that scientists often formulate theories on the basis of hunches and visions, ignoring complications and unanswered questions that often arise. He believed that the scientist is guided by a vision of a hidden reality and that the vision is dynamic. The shifting character of that vision is determined for the individual scientist by the twin characteristics of intuition and imagination.[6]

In his search for a model of a scientific manner of perceiving things in nature, Polanyi reexamined the origins of Copernicus's discovery of the solar system. Copernicus, he maintained, was possessed of a vague and extravagant vision, unsupported by the evidence, of the reality that he sought to model. Nevertheless, the general correctness of his vision was borne out by the work of others who followed, such as Kepler and Galileo. Polanyi thought that assumptions based on unaccountable clues often go into the formulation of a scientific discovery:

> And we may say this generally: Science is based on clues that have a bearing on reality:
> - These clues are not fully specifiable.
> - Nor is the process of integration which connects them fully definable.
> - And the future manifestations of the reality indicated by this coherence are inexhaustible.
>
> These three indeterminacies defeat any attempt at a strict theory of scientific validity and offer space for the powers of the imagination and intuition.[7]

According to Polanyi, intuition arises from our capacity to perceive coherences in things. The identification of a suitable problem and the search for its solution lies in an initial intuition of a coherence awaiting discovery. The quest for a solution is guided by feelings of deepening coherence, the product of a dynamic intuition. The imagination is involved in attempts to close the gap between what is known and what is thought to be the underlying reality.

Polanyi devoted a good deal of his attention to the concept of tacit knowledge.[8] This kind of implicit knowledge is gained through experiences in the world; it forms the largely unconscious basis of much of our thought and action. Tacit knowledge is not communicable; although we use it continually in everyday life, we do not, indeed cannot, communicate this knowledge in any explicit way. Polanyi uses the example of knowing how to ride a bicycle. But tacit knowledge extends to areas other than simply physical skills. For example, the expert marathon runner could never convey in explicit form all that she knows about how to run a marathon, about how to pace herself and optimize her own capacities in relationship to those against whom she is competing. The experienced synthetic organic chemist possesses a great deal of implicit knowledge of reagents, appropriate reaction conditions, and techniques for isolating products that he cannot convey explicitly. Generally, experts carrying out a complex task or solving a complex problem sense how to proceed because they are acting on the basis of acquired tacit knowledge.

Scientists, while applying their tacit knowledge in the special tasks of investigating the physical world, use the same kinds of mental processes that characterize all struggles for high intellectual attainment: "We should be glad to recognize that science has come into existence by mental endowments akin to those in which all hopes of excellence are rooted, and that science rests ultimately on such intangible powers of our mind. This recognition will help to restore legitimacy to our convictions, which the specious ideals of strict exactitude and detachment have discredited."[9]

Notice the swipe at logical empiricism in the last sentence of this quote. In his emphasis on the interaction of the individual with the physical world and the embodied nature of our conceptual frameworks, Polanyi has much in common with developmental psychologists such as Piaget and Vigotsky. Polanyi was not a cognitive scientist, but much of his thinking about the ways in which humans observe and learn is closely related to themes that have arisen in the cognitive sciences in recent decades.

Implicit Knowledge

The importance of tacit (or implicit) knowledge has become more evident in the past few decades as attempts were made to emulate the reasoning processes of experts in computer-based artificial intelligence systems. Expert system programs proved to lack the suppleness of human reasoning and decision making, no matter how intensively human experts were mined for their knowledge and reasoning processes to provide input for the programs. The missing ingredient, which failed to yield to various elicitation methods, proved to be the largely unconscious implicit knowledge that forms the basis of much of the expert's stock in trade.[10]

The processes of implicit learning and reasoning have been subjected to empirical study with human subjects. These two quotes from recent psychological literature illustrate the conclusions drawn: "Implicit knowledge results from the induction of an abstract representation of the structure that the stimulus environment displays, and this knowledge is acquired in the absence of conscious, reflective strategies to learn. . . . Implicit acquisition of complex knowledge is taken as a foundation process for the development of abstract, tacit knowledge of all kinds."[11] "Implicit learning (a) operates largely independently of awareness, (b) is subsumed by neuroanatomical structures distinct from those that serve explicit, declarative processes, (c) yields memorial representations that can be either abstract or concrete, (d) is a relatively robust system that survives psychological, psychiatric, and neuroanatomical injury, (e) shows relatively little interindividual variability, and (f) is relatively unaffected by ontogenetic factors."[12]

One of the more interesting aspects of recent psychological work on implicit knowledge is that it opens the door to inquiry about the nature of intuition:

> [Intuition] is a cognitive state that emerges under specifiable conditions, and it operates to assist an individual to make choices and to engage in particular classes of action. To have an intuitive feeling of what is right and proper, to have a vague feeling of the goal of an extended process of thought, to "get the point" without really being able to verbalize what it is that one has gotten, is to have gone through an implicit learning experience and have built up the requisite representative knowledge base to allow for such judgement.[13]

Scientific theorizing often proceeds via thought experiments, or gedankenexperiments, in which the scientist imagines states and processes that might exist in the world and analyzes them as a means of gaining insight. It is sometimes argued that the success of thought experiments demonstrates that science is at bottom a process of ratiocination, methodological reasoning in the abstract. In this way, so the argument goes, it was possible for Einstein to conceive of counterintuitive theoretical ideas that defied the evidence of the senses. But analysis of thought experiments shows that they involve processes and procedures abstracted from everyday life or from laboratory practice. They are characteristically in the form of a narrative that follows a script of events as one would experience them in the real world.[14] A key role is played by the imagination, which draws on accumulated embodied experience to construct the mental representation that constitutes the thought experiment. To illustrate, Einstein in later life described a thought experiment he developed as a sixteen-year-old schoolboy, which formed an important element in relativity theory. He imagined a light wave moving along as illustrated in figure 1.1 and himself moving in the same direction.

Suppose he were to pick a particular point on the light wave; could he catch up with it by moving fast enough? He imagined the consequences of being able to catch up with and move alongside the light wave, something that is at least conceivable in terms of Newtonian mechanics. What would the electric and

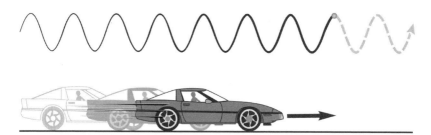

Figure 1.1. Einstein's thought experiment. What would be the implications of being able to travel at the speed of a light wave?

magnetic fields of the light wave look like? By 1905 he had come to the conclusion that an answer based on the Newtonian framework is impossible, and this line of thinking led him eventually to the special theory of relativity. It is true that the conclusions Einstein drew from his analysis of the thought experiment were counterintuitive in terms of a Newtonian conception of the world. However, the physical premises of the gedankenexperiment were extensions of Einstein's everyday physical experiences. It is a characteristic of all Einstein's reasoning from thought experiments that they are based on reasoning from deeply embodied experience. His genius lay in thinking about those experiences in novel ways.[15]

To fully understand how science works, we must look into questions of how the scientist interacts with the world. In part this involves analyzing the nature of experiment. David Gooding analyzes thought experiments this way: "Thought experiments work because they are distillations of practice, including material world experience. Experimenters move readily between representations and their objects. . . . The moral for philosophy of science of the interdependence of thought and action (or of theory and experiment) is that when you ignore one, you end up with a false view of the other and therefore, with false problems of empirical access, representation, meaning and realism."[16]

Gooding is saying that to understand the origins of scientific creativity we must know what is involved in the scientist's engagement with the world in the course of doing science. How does the scientist approach observations, reason about observational data, create and develop new theories, and communicate ideas to others? We begin by building on the premise that tacit or implicit knowledge and embodied experience play essential roles. We will see that these subconscious forms of understanding are manifested in the metaphorical character of scientific reasoning, as reflected in the language used in reasoning and communicating about science. Analysis of language provides insights into the nature of the cognitive processes used in reasoning and into the structured nature of our understanding of the physical world. We will see as we move forward that examining the roles of metaphor in scientific reasoning and communication yields powerful insights into the cognitive grounding of science.

A Look Ahead

My hope is that in the chapters that follow I will convince you of several claims about the nature of science. These are among the most important:

- Scientists understand nature largely in terms of metaphorical concepts, based on embodied understandings of how nature works. These embodied understandings derive from very basic and pervasive interactions with the physical world. They are tacit and largely common to all humans. The scientist also understands complex systems in nature in terms of conceptual frameworks derived from experiential gestalts, ways of organizing experience into a structured form. These gestalts may be drawn from the scientist's pervasive social experiences.

- The models and theories that scientists use to explain their observations are metaphorical constructs. To understand how science works and to account for its success, we have no need for the proposition that scientists have unmediated access to the world "as it really is." We have no grounds for believing that there exist objective, mind-independent truths awaiting discovery. Rather, statements we regard as truths about the world are the product of human reasoning.

I begin in chapter 2 with a brief introduction to metaphor, with examples of metaphor in language usage generally and in scientific reasoning and discourse. Among the important points of this introduction is that metaphors may be not only linguistic but also visual or tactile. The verbal metaphors used in science often are just like those used in other forms of discourse. But scientific metaphors may be more difficult to discern. In particular, we address the knotty issue of whether theories are metaphorical in nature.

Chapter 3 involves a brief introduction to the theory of conceptual metaphor. According to this theory, due largely to George Lakoff and Mark Johnson, much of our everyday thinking and reasoning about the world is metaphorical. Abstract entities such as time, inflation, love, and argument, which we cannot experience via our sensory systems, are thought about and talked about in terms of metaphors drawn from direct physical and social experiences and understandings. The theory of conceptual metaphor provides us with powerful tools for understanding how scientists reason about and communicate abstract ideas. It helps to clarify the nature of scientific creativity and enables us to relate reasoning and communication in science to other domains of thought. Finally, it also suggests constraints that must apply to scientific reasoning.

Chapters 4 and 5 take us directly into the domain of science, via a study of the history of our conceptions of atoms. Beginning with the Greeks, Western science has entertained a succession of ideas on whether the world is formed from microscopic, indestructible, individual particles called atoms. How are atoms to be understood? What metaphorical representations have been used for atoms as our knowledge of matter has grown successively more detailed? What do we really know about atoms? Are they real? Can we literally see them?

In chapter 6 we take up molecular models. The two- and three-dimensional representations are used to help understand the properties of microscopic entities not accessible to direct observation. An examination of the historical development of models and of their present-day uses illuminates the power of visual and tactile metaphors as well as their limitations.

Chapter 7 is concerned with the metaphor of protein folding, an important biological *process*, as distinct from a static structure or form. The ways in which scientists talk about processes reveal the metaphorical representations of causation that underlie understanding of events in the world. In addition, the ways in which scientists think of energy, an important abstract entity, are revealed in the verbal and visual metaphors used to describe transformations of matter, of which protein folding is an example.

Chapter 8 takes up topics from biology that involve higher levels of complexity than molecular systems. The cell is such a complex system, consisting as it does of a number of identifiable structural elements and a host of chemical substances, all behaving in a concerted fashion. To understand such systems, the biologist draws on metaphors from the social domain. One of the more interesting and productive metaphors in this arena has been that of "chaperone" proteins. The development of the chaperone metaphor, which is of recent origin, can be traced from its first applications to the broad range of phenomena that it embraces today. This example shows how scientific metaphors evolve as new data and theories, themselves prompted by application of the original metaphor, come into play. It illustrates the roles metaphors play in creative thought and communication in science.

Large, complex problems in science inevitably involve multiple metaphors operating on differing levels. The overall structure of our understanding of one such complex problem, global warming, is discussed in chapter 9. Global warming has potentially important economic, environmental, and other social implications, with resulting clashes of interests. In a case such as this, the veridical character of scientific results often is at issue. How sure are scientists of the origins of global climatic changes? How much confidence can be placed in predictions of global warming and consequent effects on climate? To place such issues in a more comprehensible framework, it is essential to understand the metaphorical underpinnings of the models used in understanding climate change. Here also we see how metaphors that move from scientific discourse to the public policy arena undergo a change in character.

Finally, in chapter 10 I have attempted to pull together the implications in prior chapters for our understanding of science. How does scientific reasoning, the analysis of observational data and theorizing, relate to other forms of intellectual activity? Does it have a special status? What can science tell us about the nature of the world around us? What kinds of truth is science capable of revealing? An understanding of the metaphorical character of scientific thought enables us to put such questions into a useful form and helps answer them.

2

INTRODUCTION TO METAPHOR

The central thesis of this book is that metaphorical reasoning is at the very core of what scientists do when they design experiments, make discoveries, formulate theories and models, and describe their results to others—in short, when they do science and communicate about it. Metaphor is a tool of great conceptual power. It enables the scientist to interpret the natural world in wonderful and productive ways. At the same time, the metaphorical reasoning that lies at the heart of scientific thought and imagination is constrained in ways that go toward defining the range and character of science.

But if metaphor is so essential to scientific reasoning, why hasn't it received more attention? Part of the answer is that philosophers have for the most part minimized the importance of metaphor in their theories of the nature of scientific activity. Volumes devoted to the structure of scientific theories or to the philosophy of science often lack even an index entry for "metaphor." One might conclude from this that metaphor isn't all that critical in science. We shall see, however, that recent advances in cognitive sciences and the emergence of new currents of philosophical thought have been changing the traditional views.

Traditional attitudes toward metaphor can be traced to the historical development of Western science. Those most responsible for developing and interpreting science in the seventeenth and eighteenth centuries had an aversion to the use of figurative language. Metaphor was viewed largely as an element of grammar and style, not as a useful device for conveying meaning. The empiricist philosophers believed that metaphor could give rise only to confusion by obscuring the categorical distinctions between words. For example, Locke expressed the philosopher's mistrust of figurative language: "All the artificial and figurative applications of Words Eloquence hath invented are for nothing else but to insinuate wrong *Ideas,* move the passions and thereby mislead the Judgment."[1]

To this day, the process of discovery in science is conceptualized by many philosophers of science, and scientists themselves, in ways that hide the full range and extent to which metaphor operates.

My intent in this book is to show that metaphor is essential to every aspect of science. It lies at the very heart of what we think of as creative science: the interactive coupling between model, theory, and observation that characterizes the formulation and testing of hypotheses and theories. None of the scientist's brilliant ideas for new experiments, no inspired interpretations of observations, nor any communications of those ideas and results to others occur without the use of metaphor.

Before we can understand how metaphor works in science, we must consider contributions from the cognitive sciences that reveal the place of metaphor in language, action, and thought. We will see that there is no separate theory of metaphor in science, as distinct from other realms of intellectual activity, nor should there be. If we are to appreciate metaphor as an essential element in the workings of science, we must understand its roles more generally in thought, language, and action.

An Introduction to Metaphors

Metaphor is traditionally defined as a form of trope, the use of a word or phrase in a figurative sense. The *Oxford English Dictionary* tells us that metaphor is "the figure of speech in which a name or descriptive term is transferred to some object different from, but analogous to, that to which it is properly applicable."[2] As we proceed, it will become evident that no single definition, such as this one, captures the full range of metaphor. Rather than concerning ourselves further with definitions, we can best begin by looking at a few examples of literary metaphor. As we do so we will see that metaphors have certain characteristics that bear on their roles in scientific reasoning and discourse. Here are the first few lines of the poem "Riverton," by Edmund Wilson:

> Here I am among elms again—ah, look
> How, high above low windows hung with white,
> Dark on white dwellings, rooted among rock,
> They rise like iron ribs that pillar night![3]

In the poem the trunks of elm trees are portrayed as girders of iron rising to the night sky. Of course trees are not literally formed of iron, nor are they rooted in rock. As I read the poem, Wilson wants to convey a sense that the trees are strong, that they in some way confer stability on the world of the town. You and I probably do not have precisely the same responses to the lines, although our readings may be generally similar. As is typical of metaphor, the meaning lies in part in what the writer has in mind and in part in the reception that the words receive in the mind of the reader.

This second example, from the poem "Continent's End," by Robinson Jeffers, is about how the sea gave birth to the life that lives on land. In the poem the poet is speaking to the sea:

> The tides are in our veins, we still mirror
> The stars, life is your child, but there is in me
> Older and harder than life and more impartial, the
> Eye that watched before there was an ocean.[4]

These lines are loaded with metaphors, such as "The tides are in our veins" and "we still mirror the stars." Beyond these individual metaphors, there is an overriding metaphorical theme, of the narrator as present at the time of creation, indeed in some sense as creator. This more extended metaphor does not reside in a few selected words but is an element of the poem in its entirety. This example illustrates the complexities that attend attempts to neatly categorize metaphor, whether in linguistic or other terms. In this connection we should note that metaphor encompasses other forms of speech and writing, such as irony, metonymy, synecdoche, and simile. The last of these deserves a brief consideration.

Similes are forms in which one thing is compared to another via the word "like." Some similes, such as "Lemons are like limes," are quite literal. Others are decidedly more metaphorical in nature, such as "His heart beat like a sledgehammer." This simile is metaphorical because sledgehammers and hearts have very few properties in common. The sense in which the heart's beat is like a sledgehammer's is just that the person is conscious of the pounding of his heart in a moment of stress as he might be of a sledgehammer being applied to rocks or the like. The beating of the heart is understood in part in terms of an entirely different entity. This particular simile, or comparison, though metaphorical, is not very complex. However, consider the opening lines from T. S. Eliot's poem "The Love Song of J. Alfred Prufrock:" "Let us go then, you and I, / When the evening is spread out against the sky / Like a patient etherised upon a table."[5]

Once again we have a simile, but in this instance there is no plausible element of similarity between the evening or evening sky and a patient under the influence of an anesthetic lying on a table. The reader of the poem must *create* a similarity by somehow attaching to the evening sky whatever moods, feelings, and thoughts might arise from thinking of a patient lying anesthetized on a table. The most interesting metaphors require the reader or hearer to do work in creating the basis of a metaphorical connection. In science, metaphors when used for the first time to characterize observations generally have this similarity-creating quality.

* * *

Figurative language abounds in poems, novels, song lyrics, and similar kinds of writing and speaking, but our everyday language is also sprinkled with metaphor:

> He is the 800-pound gorilla in that group.
>
> The president is captain of our ship of state.
>
> The Internet has become a huge shopping mall.
>
> Don't contradict him, he'll blow a fuse.

In all these cases, the sentences, if taken literally, don't make sense. We know as soon as we hear or read them that they must be understood as metaphors. The process of understanding apparently is quite facile for most people, and we generally do it unconsciously. Empirical studies have shown that it doesn't take much longer to process a metaphorical passage than a literal one unless the metaphor is inept or requires knowledge or experience that the hearer or reader lacks. But how does this processing work? We will not be reviewing the various treatments of metaphor that have held the stage at times during the past half-century or so. The interested reader can find accounts in the notes.[6] Instead, I want to use a particular way of looking at metaphor that derives from the cognitive sciences.

Let's start with a simple example from the black-and-white films of the 1930s. The reporter rushes into her editor's office and says excitedly, "Chief, this story is dynamite!" Obviously, the reporter's statement is not to be taken literally. The story is a paper document, dynamite is quite another form of chemical substance. Yet the sentence somehow makes sense; we understand immediately that the reporter is conveying something she thinks is important about her story by relating it to dynamite, a very different entity. The chief is invited to use his understanding of dynamite to gain a perspective on the reporter's story. This, in brief, is what metaphor is all about: applying information and understandings from one domain of experience, which we call the *source domain*, to enhance understanding of another domain, called the *target domain*, that is typically more abstract.

The process by which the connection is made between these two disparate domains of thought can be understood in terms of a knowledge representation scheme.[7] Suppose we list properties of dynamite that come to mind as possibly having a connection with the reporter's story. Figure 2.1 lists some of these under the heading "Source Domain," the literal domain on which the metaphor rests. Under the heading "Target Domain" we list properties that the reporter might think of her story as having. Characteristics of the source domain are associated with, or *map onto*, characteristics of the target domain. This way of representing metaphors illustrates that the target domain can be understood at least partly in terms of the properties associated with the source domain.

We have here a pretty good metaphor; when we hear it, we get it. We don't have to think about it at a conscious level. We readily pick up on the correspondences that the metaphor points to. However, we should note a very important characteristic of this and other metaphors; they *hide* many qualities of the entities being related that are not central to the metaphor. In this case, the chemical composition of dynamite, that it comes in the form of sticks in wooden boxes, must be set off by a fuse, and many other of its characteristics are not relevant to the metaphor. Only the aspects of dynamite that the listener deems important for understanding the metaphor are used. But not all listeners will have just the same background of experience and understanding of dynamite, so not everyone will interpret the metaphor in just the same way. Indeed, you may have thought of items that I did not include in the knowledge representation scheme in figure 2.1. In the process of inferring the connections between the reporter's story and dynamite, each recipient of the metaphor creates the similarity connections that give the metaphor its punch.

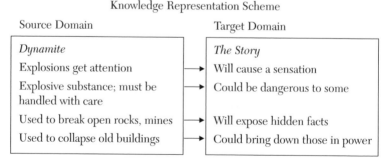

Knowledge Representation Scheme

Source Domain	Target Domain
Dynamite	*The Story*
Explosions get attention	→ Will cause a sensation
Explosive substance; must be handled with care	→ Could be dangerous to some
Used to break open rocks, mines	→ Will expose hidden facts
Used to collapse old buildings	→ Could bring down those in power

Figure 2.1. A knowledge representation scheme mapping the properties of dynamite onto a provocative newspaper story.

Channels: A Metaphor from Science

Metaphors in science can look very much like the one we have just analyzed. For example, this sentence concerning the membranes, or outer walls, of cells might appear in an introductory biology text: "Membranes contain channels that are permeable to hydrogen ions and other positive ions." In this partial description of the walls of living cells, the word *channel* is used metaphorically. The following background information will help in understanding the metaphor: The medium inside and outside cells is mostly water containing a variety of dissolved substances. Many of these substances are ionic: hydrogen, sodium, potassium, and calcium ions. This aqueous medium is called *polar;* it contrasts with the composition of the cell walls, which is largely *nonpolar.* Just as water (a polar substance) does not mix with a nonpolar material (e.g., cooking oil) in the macroscopic domain, the cell's wall does not mix with the environment inside and outside the cell in living systems. Small molecules such as oxygen diffuse rapidly through the cell wall, but polar molecules and ions normally cannot do so.

At some time in the past, someone studying the cellular concentrations of ions found that these charged, water-soluble species were moving in and out of cells, somehow passing through the cell membrane. This passage normally is very slow, but under certain conditions it can be much faster. "Channel" was chosen as the metaphorical image to represent the putative passage through which transfer occurs. The originator of the metaphor might have chosen another word, such as "tunnel" or "corridor." Or a substantially different model of the passage might have been chosen, for example, that the ions are somehow wrapped in a suitable molecular packaging and transferred through like packets. The attributes of channels in our macroscopic, everyday world, particularly their association with water, apparently mapped best onto the bare facts of ion transport as first discovered.

Once the choice of metaphor was made, its use began to *create* similarity. Those hearing the metaphor were led to think along certain lines as they conceptualized the observations. Channels have many attributes. Which of these might find a mapping in the microscopic world of the cell and its surroundings? For example, channels have boundaries. What are the boundaries of the metaphorical channels made of? Over the past four decades or so, through a large number of experimental studies, the metaphor of cell membrane channels for ion transport has become rich and detailed. For example, it has become clear that the channels consist of proteins that assume a long, cylindrical shape. There is a space down the inside of the "cylinder" through which the ions might pass. So, in a metaphorical sense, parts of the protein form the wall of the channel.

There are further entailments. In the macroscopic world, channels between bodies of water are characteristically narrow and thus closed to vessels larger than a certain size. Or the channel may have a gate, such as a drawbridge, that can be open or closed to all vessels or to particular kinds of vessels. These properties of channels in the source domain might cause the scientist to wonder whether channels in the target domain of the cell could become blocked by the action of a chemical agent or whether the channel could selectively block the passage of certain charged species. As before, we can construct a knowledge representation scheme to understand the nature of the metaphorical mapping. The source domain consists of the properties of channels as we know them in the macroscopic world. The target domain consists of the corresponding elements of the *observation domain,* as shown in figure 2.2. Notice that the source domain and observation (target) domain have nothing literally in common. The observation domain does not include actually seeing a channel.

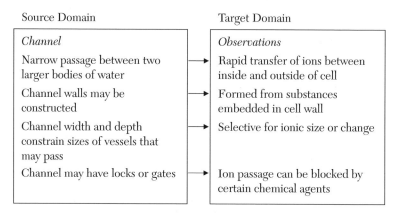

Source Domain

Target Domain

Channel	*Observations*
Narrow passage between two larger bodies of water	Rapid transfer of ions between inside and outside of cell
Channel walls may be constructed	Formed from substances embedded in cell wall
Channel width and depth constrain sizes of vessels that may pass	Selective for ionic size or change
Channel may have locks or gates	Ion passage can be blocked by certain chemical agents

Figure 2.2. A knowledge representation scheme mapping the properties of channels in the macroscopic domain onto biological channels.

The initial hypothesis that passage of substances through cell membranes involves something like channels has become over time a family of ever more detailed metaphorical models of cell wall channels of many kinds. These models have evolved in response to a succession of new results demanding new interpretations. However, the underlying idea of channel remains. Here is a passage from an introductory-level college text on biology: "Channel proteins form permanent pores, or channels, in the lipid bilayer through which certain ions can cross the membrane. Most channel proteins have a specific interior diameter and distribution of electrical charges that allows only particular ions to pass through."[8] The channels are now identified as protein in nature. To make sure that the student gets the idea, the text authors provide a simple schematic illustration (figure 2.3). This drawing is also a metaphor, or collection of meta-

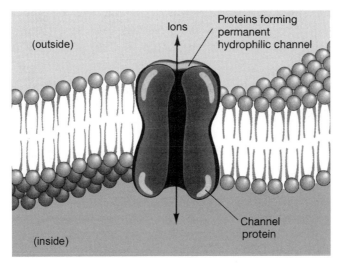

Figure 2.3. A textbook schematic illustration of a cell wall channel. (Reprinted with permission of Pearson Education, Inc., Upper Saddle River, N.J., from G. Audesirk and T. Audesirk, *Biology: Life on Earth,* 5th ed. [Upper Saddle River, N.J.: Prentice Hall, 1999], 75, fig. 5.4B. © 1999)

phors, in very much the same sense as the verbal statement: Various salient aspects of the illustration map onto the experimental data regarding channels.

Because this is our first example of a scientific metaphor, it is important to be clear about what makes the idea of channels in cell walls metaphorical. During a long period in which many studies of channels were carried out, scientists could not see channels in cell walls, even with high-powered electron microscopes. The observations that formed the target domain included data on the rates at which ions pass into or out of cells under various conditions and correlations with the presence in the cell walls of various proteins that are thought to form the "walls" of the channels. These and other observations are complemented by models of what the cell wall proteins are like in terms of shape and other properties. Scientists then constructed a model of ion passage that captures as many characteristics of the observation domain as possible. A successful model must have structure, and relationships between its parts must map onto the corresponding elements in the observation domain.[9] The model is a way for us to understand the observation domain in terms of more directly accessible experience in the macroscopic, everyday world.

When we give the name "channel" to the model, the word is being used metaphorically. The model itself is also metaphorical; it maps the properties of channels as we know them in the macroscopic domain onto a microscopic entity to which we do not have the same level of access as far as sensory data are concerned. Note that it is not just because the "channels" in cell walls are tiny and thus not directly visible that we have a metaphorical connection. It is true, as we will see in later chapters, that our only access to the structure of matter at the atomic and molecular levels is via models that are themselves metaphorical. Thus, we do not have access to something we could call a literal representation of a channel in a cell wall. But, more fundamentally, the use of "channel" here is metaphorical in the sense that we use our literal knowledge and experience of channels in the macroscopic world to characterize not only structures but also *functions* and *relationships* between entities in the microscopic domain.

Visual Metaphors

Let's turn now from verbal metaphors to a different kind of scientific metaphor. Figure 2.4 shows two representations of the substance methane, the principal component of the natural gas used to heat many homes in the United States. These are visual metaphorical representations, or models, of the methane molecule. The three-dimensional models to which these images correspond could be either visual or tactile metaphorical models. You could look at them to learn their shape and form, or you could handle them and deduce the same relevant information, other than color information, by feel.

How are these metaphors related to the examples of verbal metaphors discussed earlier? The results of experiments provide us with information about the methane molecule that maps onto the visual representations of figure 2.4. One experimental result, for example, is that there are four hydrogen atoms and one carbon atom in each molecule. Another is that all four hydrogens are equivalent; that is, they all have the same relationship to the carbon atom. A third is that the hydrogen atoms are not chemically bonded to one another but only to the carbon. Finally, experimental results from studies of carbon compounds in general indicate that the four bonds around the carbon in methane point toward the apices of a tetrahedron, a symmetrical four-sided figure.

When we look at the top model we see that it is constituted from the following basic metaphors regarding atoms and chemical bonds: "Atoms are spheres" and "Chemical bonds are rigid rods between atoms." There are many reasons, from both theory and experiment, for choosing to conceptualize atoms as spherical in shape; we will touch upon these in chapter 5. The experimental facts about chemical bonds are that two atoms connected by a bond maintain a fixed distance from one another, aside from small motions. The results of many exper-

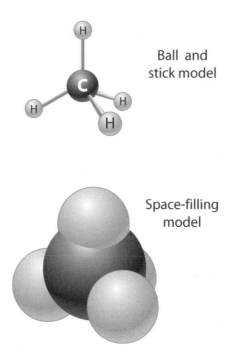

Ball and
stick model

Space-filling
model

Figure 2.4. Two representations of the methane (CH_4) molecule.

iments indicate that carbon-hydrogen bonds of a particular kind have essentially the same bond distance in many different compounds. The use of rigid rods to model the chemical bonds is thus consistent with experimental results: There is a characteristic distance between the carbon and each hydrogen. These two metaphors are combined with the experimental facts about methane to produce the model shown. The ratio of carbon to hydrogen atoms and the tetrahedral symmetry of the model are consistent with experiment.

Thus, the model captures much important information about the methane molecule. In comparing similar models of different compounds, we could make the distances between the centers of the spheres proportional to the experimentally determined distances between atoms. Coloring the atoms so that each element has a distinctive color is another way to add specificity. Hydrogen in such models typically is shown as white and carbon as black. Thus, anyone who knows the color scheme can look at the model and know that it is intended to represent methane and not silane, which is like methane but has a silicon atom in place of the central carbon atom. In silane the central atom would be shown as a color other than black.

The ball-and-stick model accounts for important aspects of the methane molecule, but it is inconsistent with others. Experimental evidence indicates that atoms are not like hard spheres but more like rubber balls that grow softer at their outer edges. A hard rubber ball on the inside with a Nerf ball–like exterior would serve as an appropriate metaphor. When atoms collide with one another, there is a bit of give as the softer outer layer is penetrated, but then they repel one another. Data from experiments permit one to estimate (using a model of some kind, of course) the distances between centers when the colliding atoms are as close as they typically come. These distances are distinctly larger than the distances between the centers of bonded atoms. In other words, the atom seems to have one characteristic radius when it forms bonds and a larger radius when it encounters atoms to which it is not bonded. That larger radius is called the van der Waals radius, in honor of the Dutch scientist Johannes van der Waals, who first proposed these ideas.

The space-filling model shown in the lower half of figure 2.4 is intended to show the atoms as though their radii were the van der Waals radii. In this model we can't see the chemical bonds. The model as a whole looks like something closer to a ball. Still, we can see that there are four hydrogen atoms and one central carbon atom. The four hydrogen atoms are still rigidly in place, distributed symmetrically around the carbon. The space-filling model offers a new perspective: This is what methane looks like to anything it collides with. It gives us a picture of the outer surface of the molecule. The verbal metaphors embodied by this model might go something like this: "The methane molecule is a ball-like entity" and "The methane molecule has a continuous, hard surface."

It is difficult to discern much about the bond between hydrogen and carbon from the space-filling model, although in fact the distances between atomic centers are accurate in relationship to the estimated van der Waals radii. We see again that although metaphors emphasize one set of features of the domain onto which they map, they hide others. The ball-and-stick model tells us nothing about the effective sizes of atoms relative to one another, nor of the existence of van der Waals radii. The space-filling model provides this information admirably, but at the expense of disguising the chemical bonds between carbon and hydrogen.

Models of the sort we have just considered, and much more elaborate ones born of various kinds of computer-based representations, are of great importance in the chemical and biological sciences. They have become so commonplace in scientific explanation, and they can be so beautiful, that it is easy to succumb to the idea that they are literal descriptions. We will take up a more detailed consideration of molecular models in chapter 6.

Models as Metaphors

We have now seen examples of what are commonly called models. The drawings of figure 2.4 are visual models of a molecule. Their three-dimensional analogs are visual or tactile models of the same molecule. The verbal description of ion channels in cell membranes, and the drawing in figure 2.3, are both models; one is verbal, the other visual. In all these examples, the metaphors map onto several features of the observation entity they are intended to represent. In some circles it is deemed important to distinguish metaphors from models; the former term is restricted to linguistic forms, whereas models are *entities* of one kind or another.[10] So one might say that a model stands apart from the metaphors it represents. I prefer not to complicate matters in this way; I don't think we can get into much trouble by treating models as extended metaphors. Typically, a model has structure and multiple elements that map systematically onto the target domain. Furthermore, as we saw in the case of the "channel" metaphor, models give rise to metaphorical entailments.

One of the most active fields of study in molecular biology today deals with the ways in which proteins change their shapes and configurations in solution. Proteins are long chainlike molecules. Under appropriate conditions most proteins that are active in biological systems coil up and rearrange lengths of the chains so as to assume a characteristic shape. This process was called "folding" because an analogy was seen between the change the protein undergoes and the folding of objects in the macroscopic everyday world, such as napkins or card table chairs. We examine the "protein folding" metaphor in a later chapter.

Is "protein folding" a model? Some would argue that the term was first used merely to give a name to a new phenomenon—an example of catachresis, defined as a deliberately paradoxical figure of speech or a strained use of a word. Catachresis is one of the means by which newly discovered things get named. But "protein folding" is more than just a name for a process. As a metaphorical expression it invites us to probe the cross-domain mapping between the literal, everyday act of folding and the changes that occur in a protein as it undergoes the transition we call folding. Thus, the act of naming the process "folding" *creates* similarities. The metaphorical expression suggests the kinds of questions that lead to a more extended model: What constitutes folding in a metaphorical sense? In everyday acts of folding, parts of the folded object are brought into contact, and the folded object is more compact overall. What does this mean for protein folding? How does folding occur? How have the properties of the protein changed when folding has occurred? Questions such as these helped to direct the course of further experimental work. Because it was a productive metaphorical description, "protein folding" has evolved into several detailed (and competing) models that map onto increasingly detailed experimental results.

We will see in later chapters that metaphors in science evolve in a characteristic way. An initial metaphor may be highly suggestive but devoid of much content. However, if it is well chosen, it suggests experiments that lead to elaboration of the metaphor.[11] There is an interaction between the metaphorical frame of thought and the literal observation system in which experiments are performed.[12] An apt metaphor suggests directions for experiment. The results of experiments in turn are interpreted in terms of an elaborated, improved metaphor or even a new one. At some stage in this evolutionary process the initial metaphor has acquired sufficient complexity to be considered a model.

In summary, models are extended metaphors that have the potential to guide thinking about a system under investigation, suggesting new directions for research. They also pose a danger: Attachment to a particular model can inhibit thinking in other, possibly more productive ways about the system being studied.

Hypotheses and Theories as Metaphors

There has been a huge amount of writing and discussion in recent decades about the structure and status of scientific theories. This book is not primarily about those matters, but our exploration of the roles of metaphor forces us to examine the processes of theory formation and their cognitive functions. I want to convince you that the hypotheses and theories used in science are metaphorical in nature. I should say at the outset that this is a proposition that many scientists and science philosophers are not prepared to accept. For example, Arthur Miller, in his recent book *Insights of Genius*, advances a commonly held view; he agrees that metaphor plays a central role in scientific creativity. With respect to the nature of theories, however, he takes precisely the opposite view to mine: "Metaphors are an essential part of scientific creativity because they provide a means for seeking literal descriptions of the world about us. *Those literal descriptions are scientific theories.*"[13]

Over the past fifty years, several philosophers of science and notable scientists have argued for the metaphorical character of scientific explanation. Mary Hesse wrote in 1966 that "the deductive model of explanation should be *modified and supplemented* by a view of theoretical explanation as metaphoric redescription of the domain of the explanandum [the observational data to be explained]."[14] In a related vein, Michael Bradie suggests that the interpretation of theories is accomplished by means of metaphor.[15]

In his recent book *Real Science*, the distinguished physicist John Ziman recognizes the metaphorical character of scientific reasoning and the theories that result:

> Even the most austerely "scientific" models operate through *analogy* and *metaphor*. The Rutherford-Bohr model depicts a hydrogen atom as a miniature solar system. Darwin's concept of "natural selection" is analogous to the "artificial selection" practiced by animal breeders. "Plate tectonics" is about thin, flat rigid areas of "crust" floating on a highly viscous but fluid "mantle." Linguists talk of the "brain mechanism" by which grammatical language is generated, as so on.
>
> Scientific theories are unavoidably metaphorical. Indeed, how could even the most original scientist construct, make sense of, or communicate to others a theoretical model that did not incorporate a number of familiar components?[16]

Colin Turbayne, in his book *The Myth of Metaphor,* argues that scientific concepts are inevitably metaphoric, but this fact has not been recognized in the evolution of Western science: "There is a difference between using a metaphor and being used by it, between using a model and mistaking the model for the thing modeled. The one is to make believe that something is the case, the other is to believe it."[17] By way of illustration, Turbayne points to the legacy of both Descartes and Newton. Bemused by the success of their attempts to model the mechanical properties of the world, both came to think of the world literally as a giant machine with particular properties.

Newton and his followers believed in the reality of the absolutely linear three-dimensional space that formed the framework of his mechanics and in the uniform, linear, and independent flow of time. John Barrow put it this way: "Newton's first universal description of a natural phenomenon was heralded by the supporters of his methods as the uncovering of a fundamental law of Nature—thinking one of the thoughts of God after him. . . . Too remarkable to be viewed in any spirit other than the realist's, it became the mainspring of the new mechanical paradigm of a clockwork Universe running to definite mathematical rules."[18]

Newton's paradigm, which seems so firmly grounded in sensory experience, proved to be not a literal account of reality but a metaphor. Einstein's theory of relativity did more than merely correct Newtonian mechanics for effects seen only under special conditions. More fundamentally, it forced an entirely new conception of the relationship between space and time and did away with the concept of force at a distance that seemed so utterly literal in Newtonian mechanics.

Contemporary science has inherited, mainly as a legacy of the seventeenth century, the idea of laws of nature as universal truths, independent of human beliefs but discoverable by humans. Newton thought in these terms because he saw the laws of nature as prescriptions from God. He believed that if we were clever enough, we could think the thoughts of God, as Barrow put it so well.

Descartes's and Newton's views, in secularized form, have continued to influence how we think of scientific laws. However, there are good grounds on which to question the usefulness of such an interpretation of scientific understanding. Nancy Cartwright and Ronald Giere, in writing recently on this subject, call attention to the discordances between such absolutist claims for scientific truth and the actual practices of science.[19]

Thomas Kuhn, famous for *The Structure of Scientific Revolutions*, wrote on aspects of the role of metaphor in science. He suggests that models such as the solar system model of the atom play important roles in guiding theory development and explanation, even after the theory seems to have advanced beyond the model. This may happen because the underlying conceptual framework persists. (We'll deal with this particular case when we come to looking at modern conceptions and theories of the atom.) Kuhn claims that there is a coupling between the language of metaphors and the theories based on them: "Metaphor plays an essential role in establishing links between scientific language and the world. Those links are not, however, given once and for all. Theory change, in particular, is accompanied by a change in some of the relevant metaphors and in the corresponding parts of the network of similarities through which terms attach to nature."[20]

We see in this quote an idea common to much writing about the roles of metaphor in science: The metaphorical character of models is granted, but they are viewed as distinct from hypotheses and theories. Of course, a theory cannot be entirely abstract; it must be *about* something, must refer to some set of observations. The logical empiricists wanted to say that a good theory provides a description of a set of observed phenomena, with terms of the theory corresponding to elements of the observation domain. It seemed appropriate in their way of looking at things to view theories as literal descriptions. Because of the direct correspondence of elements of the theory to things in the world, a model was deemed superfluous. The trouble with this limited view of theories is that it fails to capture how most science takes place. Most typically, the scientist attempts to understand a set of observations in terms of a model of the system under study. The model may be one that has been used before, perhaps modified to make it consistent with new observations. However, when the observations are of a new kind or made on a system never before studied, an entirely new model may be needed.

How is the model distinct from the theory? Consider the channel model of ion transport discussed earlier. The elements of the model don't correspond literally to elements of the observation domain. Rather, the model has structural and causal features that map onto the observations. For example, a particular channel model may have the feature that ions of the presumed size and charge of calcium ions move preferentially through the channel under the influence of an electrical potential. This feature of the model maps onto observations of calcium ion transport into and out of cells. The model and the story of how it matches, or accounts for, the observed measurements of ion movement in and out of cells constitutes a theory of ion transport. The theory might be made quantitative with mathematics. If it were, the terms of the mathematical expressions, such as diffusion rates, channel diameter, and interactions between charged species would match elements of the model. The values of adjustable terms would be chosen to afford the best match of theory prediction with experiment. In other words, the mathematical representation, which some would call a theory of ion transport across membranes, is not really distinct from the model; it is an expression of it in mathematical language. The theory is just as thoroughly metaphorical as the model.

The question of whether theories and hypotheses are metaphorical or literal is important, but we can put it aside for now. We will be in a much better position to evaluate conflicting points of view after examining the roles of scientific metaphors in many different contexts.

Summary

This brief introduction to metaphor has identified themes that will recur in one form or another throughout the book. These are among the most important:

- Metaphors can be thought of as mappings from a source domain of literal, everyday experience to a target domain, with the aim of enlarging and enhancing understanding of that target domain. We use understandings from the domain of direct physical and social experiences to structure our understanding of a more abstract domain.
- A given metaphor highlights certain features of the source domain and hides others, depending on the intent of the author. Often, however, some of the hidden elements are implied by the author or are inferred by the recipient, depending on context. It is just these implications that make metaphor a powerfully creative force in scientific reasoning.
- Although metaphors invite comparisons of two disparate things, the more interesting metaphors do more than this. They stimulate *creation* of similarities between the source and target domains, such that the target domain is seen in an entirely new light.
- Metaphors in science serve an explanatory role and are a stimulus to new experiments. They may be very simple and evocative initially, then grow more detailed as research findings support or disconfirm inferences drawn from the initial metaphor.

• Models, which are extended metaphors, give rise to metaphorical entailments, which influence the ways in which the model is understood and applied. Models commonly form the basis for theory formation.

In the chapters ahead we will build on these themes to show that metaphor is a key to understanding the most important aspects of scientific activity. We will see that metaphors serve social roles in science, such as promoting a particular idea or staking a priority claim. We turn in the next chapter to a brief look at the theory of conceptual metaphor, which provides valuable tools for detailed examination of important metaphors in the chapters that follow.

In the long history of writings on metaphor, beginning with Plato and Aristotle and extending through Richards and Black, metaphor has been identified as the use of words and expressions outside their normal, conventional meanings. We saw in chapter 2 examples of metaphor in literary contexts, in everyday figurative language, and in scientific communication. But we saw also that metaphor encompasses more than just linguistic devices; physical models and drawings may also be metaphorical.

Within the domain of verbal metaphors, the distinctions between so-called literal and metaphorical uses of language have always been uncertain. Theories of metaphor have come and gone without shedding much light on their roles in reasoning and communication. Recently, the study of metaphor has moved from primarily literary and philosophical territory to the realms of psychology, linguistics, and other cognitive sciences. As cognitive scientists have learned more about human conceptual systems, the essential roles played by metaphorical thought have become more evident. Many of the entities that we want to think about and talk about, such as love, time, or the meanings of scientific observations, are abstract concepts. To convey ideas about these abstract entities, we call upon language and conceptions that we normally use in speaking and thinking about more concrete experiences. David Rummelhart points by way of example to our conceptualizations of mind:

> Nearly always, when we talk about abstract concepts, we choose language drawn from one or another concrete domain. A good example of this is our talk about the mind. Here we use the spatial

model to talk about things that are clearly nonspatial in character. We have things "in" our minds, "on" our minds, "in the back corners of" our minds. We "put things out" of our minds, things "pass through" our minds, we "call things to mind," and so on. It is quite possible that our primary method of understanding nonsensory concepts is through analogy with concrete experiential situations.[1]

That is, we understand abstract concepts by metaphorical mappings from source domains based on direct physical and social experiences. In this chapter we examine the basis of such metaphorical mappings.

Conceptual Metaphors

The spatial metaphor for mind is but one example of the general observation that we talk about abstract concepts by using language drawn from concrete domains. Such use of language is metaphorical, but not in the sense used in classical theories of language, which concerned themselves with novel constructions, in which words are not used in their ordinary senses. The new view of metaphor asserts that our everyday language is replete with metaphors that we use without being conscious of their metaphorical character. These are called *conventional metaphors* or, preferably, *conceptual metaphors,* to distinguish them from the novel constructions found in fiction, poetry, and scientific theories. George Lakoff asserts that conceptual metaphors are not simply matters of language: "The locus of metaphor is not in language at all, but in the way we conceptualize one mental domain in terms of another. The general theory of metaphor is given by characterizing such cross-domain mappings. And in the process, everyday abstract concepts like time, states, change, causation, and purpose also turn out to be metaphorical."[2] Note that Lakoff here refers to metaphor as a conceptualization process. The linguistic expression that may result from this cross-domain mapping is a surface manifestation of a more fundamental and deeper matter of thought.

In 1980 George Lakoff and Mark Johnson published the book *Metaphors We Live By,* in which they described the theory of conceptual metaphor, citing a host of examples drawn from a wide range of human experience.[3] These authors have further developed their theory of metaphor in more recent books.[4] Although the literature of the field has grown enormously since its appearance, *Metaphors We Live By* remains a widely read and quoted statement of conceptual metaphor theory. In what follows I will attempt to extract the most important aspects of their ideas for our purposes in understanding the roles of metaphor in science.[5]

We carry around in our heads a large array of concepts that govern our thought processes and everyday functioning. They determine how we perceive the world, even what we perceive, how we navigate through our daily lives, and how we relate to others. Many of the most important entities for which we must have conceptual representations, such as time, love, and inflation, are abstract. To conceptualize such abstract domains of thought we relate them to more concrete concepts with which we have direct experience. We do this by mapping across domains, making connections between the elements of the more abstract conceptual domain and corresponding elements of the more concrete one. We saw a few simple examples of mappings of this kind in chapter 2. In those cases, we readily identified the mapping as metaphorical. But cross-domain mappings show up in our language use as what are often called conventional metaphors. Lakoff and Johnson provide evidence that these metaphors are based on underlying conceptual structures that derive from our embodied experiences with the world. It is for this reason that they are properly called conceptual metaphors.

To illustrate how this works, let's consider the metaphor "An argument is a construction." *Argument* is used in the sense of reason or reasons offered in support of a proposition or theory. (Most commonly, the metaphor notation is a shorthand of the form "Target domain is source domain." The notation "An argument is a construction" denotes a set of correspondences between our understanding of certain properties of arguments and our understanding of analogous properties of constructed objects. It is the set of correspondences that makes up the mapping from the source to target domain.) Here are several sentences that illustrate this general metaphor, the first two from Lakoff and Johnson:[6]

> He is trying to *buttress* his argument with a lot of irrelevant facts, but it is still so *shaky* that it will easily *fall apart* under criticism.
>
> With the *groundwork* you've got, you can *construct* a pretty *strong* argument.
>
> If his key assumption is disproved, his argument will *collapse*.
>
> Smith's argument is *built on* two key sets of experimental results.
>
> These latest results *undermine* most of his argument.

The italicized words in each case reveal the metaphorical conception of an argument as a construction.

There are other commonly used metaphors for argument, such as "An argument is a journey":

> Let's see how *far* this line of argument can *carry* us.
>
> We've *come to* an important *point* in the overall argument.
>
> I think she's correct, but her argument is pretty *roundabout*.

> As we *move* a little further *along* in the argument, you'll see what I mean.
>
> Look, here's where his argument *goes* completely *off the track*!

These examples show how we understand the making and presenting of arguments in terms of a construction or a journey. Constructions and journeys are familiar aspects of our lives. Of course, arguments are not physical objects or journeys, but some of the properties of arguments map onto corresponding properties of constructions or journeys. If we can conceive of arguments as metaphorically related to constructions or journeys, then the activity of making an argument can be conveyed metaphorically, and we can talk about arguments in terms of the metaphorical relationships to constructions or journeys. Notice that we have not specified what sort of construction is involved or by what means of transport the journey is undertaken. "Construction" and "journey" are called superordinate categories; "construction" includes any number of more basic-level terms for buildings, such as "house," "church," and "office building"; a "journey" might be undertaken by any vehicular means, such as car, train, or airplane. In some cases the metaphor may be sharpened by a choice of one particular basic-level term, but more commonly the more general superordinate level of reference is used.[7]

* * *

We can see from these examples that words used to characterize constructions or journeys are regularly used to talk about arguments. Words such as "build," "buttress," and "undermine" reveal how we conceptualize arguments as structures. Similarly, such phrases as "come to," "move along," and "goes off the track" demonstrate our conceptualization of argument as a journey. In both instances these metaphors are systematic; the general metaphor carries with it a host of potential entailments that could follow from the core metaphor. These conceptualizations affect the ways in which we form arguments and understand them. In "formulating," "assembling," and "constructing" an argument or thesis, we proceed as though we were putting together a construction. Alternatively we might use the "journey" metaphor, thinking of where we need to get to, the stages along the way, what it will take to get us to each stage, and so on. Beyond this, we might use novel metaphors based on these conventional metaphors to make a point more dramatically, convey irony, or address some other aspect of the general theme. For example,

> Smith's argument is basically an attempt to shore up a decrepit structure.
>
> I followed his argument to the end, but the trip wasn't worth it.
>
> This theory is OK, but it's one of the wings, not the whole cathedral.

Ontological Metaphors

The metaphor "An argument is a construction" is an example of what Lakoff and Johnson call an *ontological* metaphor, in which abstract concepts, such as ideas, events, and activities, are thought of as entities and substances. The abstract notion of argument is understood and talked about in terms of physical construction. Ontological metaphors abound in our thinking and use of language. Here are several examples:

Pornography makes me sick.

The Federal Reserve is always wary of high inflation.

My mind just isn't operating at full capacity today.

In dealing with these people, weigh carefully everything they say.

In these examples, four abstract concepts are spoken of as though they had the properties of tangible things. Pornography is seen metaphorically as a poison, inflation as a potential adversary to be carefully watched, the mind as a kind of machine, and statements as objects with weight.

Time

An important example of ontological metaphors is the way in which we talk about the abstract idea of time. Because it is so central to our lives, no one metaphorical representation of time suffices to express the many ways in which it must be conceptualized. Among the many metaphorical mappings for time, one of the most general is "Time is a resource." A subset of this is "Time is money." Here, from Lakoff and Johnson, are some of the manifestations of this conventional metaphor in contemporary English:

You're wasting my time.

This gadget will save you hours.

I don't have the time to give you.

How do you spend your time these days?

That flat tire cost me an hour.

I've invested a lot of time in her.

You're running out of time.

He's living on borrowed time.

I lost a lot of time when I got sick.

Thank you for your time.[8]

We don't need to look far for the origins of this conventional metaphor in our social experience. Workers' pay is expressed as so much per hour; we rent things by the month or week and pay for services on the basis of so much per unit of time. We have annual budgets, quarterly earning reports, daily hotel room rates, and so on. The metaphor "Time is money" is part of the larger metaphor "Time is a resource," something that can be spent, saved, wasted, sold, or squandered. The major point for the present is that the abstract quantity, time, is conceptualized in terms of entities that we deal with in everyday life. Metaphorically speaking, time is an entity, a "thing."

The *passage* of time is conveyed using metaphors related to flow or movement. So we say, "time flies" or "with the passage of time." These concepts of time imply a spatial representation of some kind. Frederick Waismann, one of the early-twentieth-century Vienna Circle (logical positivist) philosophers, had these wry comments to offer on the perplexities of the quest for an understanding of time:

> "Time flows" we say—a natural and innocent expression, and yet one pregnant with danger. It flows "equably," in Newton's phrase, at an even rate. What can this mean? When something moves, it moves with a definite speed (and speed means: rate of change in time). To ask with what speed time moves, i.e., to ask how quickly time changes in time, is to ask the unaskable. It also flows, again in Newton's phrase, "without relation to anything external." How are we to figure that? Would it flow on irrespective of what happens in the world?[9]

Time often is conceived of as a river, with the future flowing toward the observer and the past receding in the opposite direction. The general spatial metaphor for time is exemplified in many everyday expressions, such as

I'll see you *at* four this afternoon.

I'll see you *in* three days.

Meteor showers occur at several times *throughout* the year.

In the first of these examples, a specific time is a location. In the other two, it is conceptualized as a container.

Several interesting entailments follow from the general conceptualization of time in terms of space.[10] The mapping from the source domain of space to the target domain in this case means, as we have just seen, that specific times are entities. Future times are in front of the observer, past times are behind the observer, and the passage of time is continuous and one-dimensional. The metaphor "Time passing is motion" has two special cases. In one, the observer is fixed, and times are entities moving with respect to the observer. Time has a velocity relative to the observer. On the other hand, times may be imagined as fixed locations, and the observer moves with respect to them.

These ways of conceptualizing time show up in everyday language when we say, for example,

> In the coming months . . .
>
> I'm looking ahead to summer vacation.
>
> The time has long since passed when . . .
>
> She's facing the future with optimism.
>
> I can't believe how quickly the time has passed!

In science the conceptualization of time as a spatial entity is reflected in the general mapping "Time is length." For example, data may be collected at specific time intervals over a period of time and the results displayed as a two-dimensional graph in which the measured quantities are displayed along one axis and time along the other. The length of the time axis represents the time elapsed from some reference starting time.

In Newtonian mechanics, time is a separate dimension, independent of the three dimensions of free space. Newton conceived of time as flowing continuously and endlessly, independent of occurrences in the spatial world. In his special theory of relativity Einstein introduced a new way of thinking about the temporal domain. He asked what is meant by the statement that two events are simultaneous. In addressing this question, the special theory of relativity postulates that space and time are not separate, independent entities but form a four-dimensional continuum called space-time. In relativity theory, "Time is length" is present as an explicit feature.

Evidently, we find it necessary to use various metaphors to reason and communicate about time, and some of them may be mutually exclusive. A given conceptual metaphor may be applicable in one situation but not in another. Time is conceptualized as an entity, but what kind of entity? Clearly, it is conceptualized as different things in these five examples:

> No matter how much time passes, we will remember this day.
>
> I don't want to spend my time that way.
>
> Time is on our side in this affair.
>
> Hurry, we're running out of time!
>
> That was the longest afternoon of my life.

The fact that an abstract entity such as time is conceptualized in many different ways is consistent with the theory of conceptual metaphor. Because we have many different experiences of time, we need differing metaphorical concepts to structure those experiences.

Orientational Metaphors

A major tenet of the theory of conceptual metaphor is that we understand abstract concepts in terms of concrete experiences and feelings. Most powerful in this regard are our direct physical experiences of living on Earth. Beginning at some early stage in development, we begin to experience the world outside our bodies. We learn to distinguish this outside world from the one within ourselves. We learn about the force of gravity and other forces, distances, depths, balance, and symmetries. We learn that when an opaque object is placed between us and another object, it obstructs our view of the farther object. All the lessons learned in development become part of the way we conceptualize the physical world.[11] We acquire what Mark Johnson calls image schemata.[12] Image schemata are not pictures, ready for calling up when we need to understand a particular abstract concept. Rather, they are structures based on bodily experience that organize the conceptual system at a more general, abstract level than any particular image. We rely on image schemata when we attempt to conceptualize more abstract ideas. This process is at work in our use of orientational metaphors, based largely on spatial orientations such as up-down, in-out, front-back, on-off, and deep-shallow.

Use of the vertical dimension in orientational metaphors is especially prevalent. The many examples of this conventional metaphor type are based on fundamental physical experiences with verticality, which arise because we and the objects with which we deal in our lives are subject to the force of gravity. The metaphor "More is up" is one of the most prevalent examples of an orientational metaphor:

> Sales of handheld computers keep going up.
>
> His temperature was high when he had the flu.
>
> Underage drinking is a problem in this town.
>
> Neurons that are not being stimulated fire at a low rep rate.
>
> Inflation dropped to a new low last quarter.
>
> Enrollments are down for the third year in a row.

The experiential basis for this metaphor is easy to discern. When we pour liquid into a glass, the level of fluid rises; as we shovel dirt on a pile, it grows taller, and so on.

We also apply the up-down orientation to abstract entities that don't involve quantity. For example, in the social domain, "High status is up; low status is down":

> He'll rise to the top of the organization.

Women's careers in corporations are impeded by the glass ceiling.

He's at the peak of his career.

Everyone in this neighborhood is upwardly mobile.

The grounding of such metaphors is not the same sort of direct physical experience as "More is up." Rather, there is a general understanding that in the social domain "Better is up." Where does such an understanding arise? There are plenty of historical origins, but by way of contemporary example we need only ask where the rich and famous live. Which pieces of residential real estate in San Francisco are the priciest? Those on the hills, with the best views, of course. And where do the privileged live in cities such as New York? On the top floors of the tallest residential buildings, of course. Orientational metaphors that are strongly cultural in content form an internally consistent set with those that emerge most directly from our physical experience. The up-down orientational metaphor can apply to situations that contain both physical and cultural elements, such as

He's at the peak of health.

She came down with pneumonia.

Here good health is associated with "up," in part because of the general metaphor that "Better is up" and perhaps also because when we are well we are on our feet, and when we are ill we are more likely to be lying down.

Other orientational metaphors are obviously cultural in origin:

He's one of the higher-ranking officials in the agency.

These people have very high standards.

I tried to raise the level of the discussion.

Whether the experience on which an orientational metaphor is based is directly emergent physical experience or one drawn from the social domain, the core metaphorical framework is the same in all of them. There is only one verticality concept "up." We apply it differently, depending on the kind of experience on which we base the metaphor.

Container Metaphors

Aside from the up-down orientational metaphor, one of the most pervasive metaphors involves the in-out orientation. Our direct experience with containers begins with our awareness of our own bodies as having a discrete boundary. We take things into our bodies, and things come out of them. We also learn about containers of various kinds and learn that we can put things into and take things out of them. But much of what we encounter in the physical world does not have obvious boundaries. For example, a clearing in the woods, a mountain range, or a subdivision may not have sharply delineated boundaries. It is helpful in talking about such entities to treat them as though they were containers, with discrete boundaries. Thus we have expressions such as

> I picked these flowers *in* the field back of the barn
>
> She lives *in* southwest Chicago.
>
> Kemper Creek has its origin *in* the Pine Barrens.

In these examples of orientational metaphors, land areas are conceptualized as containers. More abstract entities such as thoughts, feelings, events, actions, activities, and states are conceptualized as objects. For example, an event is conceptualized as an object that may have the properties of a container. Lakoff and Johnson give these examples to illustrate:

> Are you going to the race on Sunday? (race as OBJECT)
>
> Are you running in the race on Sunday? (race as CONTAINER OBJECT)
>
> Did you see the race last Sunday? (race as OBJECT)
>
> Halfway into the race I ran out of energy. (race as CONTAINER OBJECT)
>
> He's out of the race now. (race as CONTAINER OBJECT)[13]

Ontological and orientational metaphors can be quite complex; they allow us to elaborate our ideas about nonphysical entities. To illustrate, aspects of society can be expressed via the metaphor "Society is a container":

> There is *room* for every point of view *in* this society.
>
> Many of the poor have basically *dropped out of* society.
>
> We believe in an *open* society.

The Grounding of Metaphors

Lakoff and Johnson claim that most of our normal conceptual system is metaphorically structured. Most of the concepts to which we have frequent and varied reference are at least partially understood in terms of metaphorical reference to other, more concrete concepts. This means in turn that our conceptual system must be grounded in a core set of direct experiences that are not themselves dependent on a metaphorical relationship. The core concepts must be those that arise from our most ubiquitous physical experiences, such as verticality, space, and vision. Such concepts are called directly emergent. They are grounded in our daily activities, in continually performed motor functions and regularly perceived events.

Some of the broadest and most provocative conclusions drawn by Lakoff and Johnson arise from their assertion that no experience is purely physical in character:

> What we call "direct physical experience" is never merely a matter of having a body of a certain sort; rather, every experience takes place within a vast background of cultural presuppositions. It can be misleading, therefore, to speak of direct physical experience as though there were some core of immediate experience which we then "interpret" in terms of our conceptual system. Cultural assumptions, values, and attitudes are not a conceptual overlay which we may or may not place upon experience as we choose. It would be more correct to say that all experience is cultural through and through, that we experience our "world" in such a way that our culture is already present in the very experience itself.[14]

Physical experiences, such as sitting, are to be distinguished from more cultural experiences, such as participating in a graduation ceremony. An important aspect of the issue of grounding is that physical experience is not somehow more basic than emotional, intellectual, cultural, or other kinds of experience. However, it is more clearly delineated because it follows directly from the workings of our sensory system. Experiences in less clearly delineated domains may be as powerfully felt as any physical experiences, but we express them in terms of the more clearly delineated physical domain. Consider the following related examples:

Sally is in the shower. (shower as container)

Sally is in New Orleans. (city as container)

Sally is in the Friday discussion group. (social group as container)

Sally is in a bad mood. (emotional state as container)

These four cases refer to equally valid and basic kinds of experiences. It is tempting to think of each sentence as a literal statement. But Sally has a different kind of experience in each case: simple physical location, location in a complex physical and cultural milieu, membership in a social organization, and being in a particular emotional state. The statements thus range from literal to metaphorical applications of the container concept.

Experiential Gestalts

As we have seen, the theory of conceptual metaphor is based on the idea that our thinking, language, and actions are based in large measure on a metaphorically structured conceptual system. We conceptualize and talk about abstract ideas in terms of more concrete ones. This is possible because we are able to map (i.e., identify a correspondence between) the elements of a concretely based concept and those of a more abstract one. We map the elements of the more concretely based concept (which forms the source domain) onto corresponding elements of the more abstract concept (which forms the target domain). But what are these elements for which correspondences are made? In some cases the concept is very simple, grounded in everyday physical experience (e.g., an orientational concept such as verticality). On the other hand, when the source domain involves activities and relationships, a complex set of related elements is involved.

To illustrate, consider the metaphor discussed earlier, "An argument is a journey." The source domain in this case is the concept "journey." In using this metaphor to think and talk about an argument or theory, we call on our experiences with journeys, which are complex activities. Based on experiences with journeys of various kinds, we conceptualize a prototypical journey as having several elements, including the following:

> Journeys are undertaken by people.
>
> Each journey has a starting point and destination.
>
> A successful journey may require advance planning.
>
> Each journey follows a route taking us from our starting point to the destination.
>
> Some means of conveyance (walking, car, rail, aircraft) takes us on our journey.
>
> Some parts of the journey may be more difficult than others.
>
> We might lose our way and go in a direction that will not lead us to our destination.

These elements, and others that one might think of, together constitute our concept of a journey. It is not a single idea but a collection of related ideas, forming a structured, multidimensional whole. Lakoff and Johnson call this collection an experiential gestalt[15] or image schema. The various dimensions that make up the whole are categories that emerge naturally from our experiences. In the metaphor "An argument is a journey," the elements of the gestalt or image schema for journey are mapped onto elements of the process of formulating and presenting an argument or theory. Here, for example, are statements about arguments and theories that more or less map onto the ones listed earlier for journeys:

> An argument is formulated by someone and followed by others.
>
> An argument has a starting point and an intended point of completion.
>
> To successfully formulate an argument may entail research into background materials and methods.

A complex argument or theory involves a progression from one step to another, in an ordered sequence.

There may be choices of theoretical methods or background materials to use in formulating and advancing the argument.

Certain steps in an argument or theory may be difficult to understand or prove.

It is possible that, by proceeding along a certain direction in terms of the model used or theoretical tools chosen, the argument will come to a logical cul de sac.

Many "literal" expressions we use in talking about arguments and theories involve imagery associated with journeys:

The first steps in the proof . . .

At this stage in the argument . . .

When I use renormalization methods here I come up against a dead end . . .

This is no mere happenstance; our understanding of argument or theory is structured metaphorically, partially in terms of the metaphor "An argument is a journey." This particular metaphorical mapping is grounded in our experience and finds its expression in common phrasings such as those illustrated earlier. We regard these expressions as literal language in the sense that our use of them is conventional and we are largely unconscious of any metaphorical character. Nevertheless, our concept of argument is shaped by one conceptual metaphor or another, such as "An argument is a journey" or "An argument is a construction." In this way, these metaphorical mappings determine how we think about arguments and theories and how we act in formulating or evaluating them.

It is worth repeating a point made earlier: Although we might attempt to represent any given gestalt or image schema as a list of propositional phrases, such as those given above for "journey," it does not exist simply as such. Rather, it corresponds to something much richer and more complex. Johnson calls them "structures of embodied understanding." He takes "understanding" to involve "*our whole being*—our bodily capacities and skills, our values, our moods and attitudes, our entire cultural tradition, the way in which we are bound up with a linguistic community, our aesthetic sensibilities, and so forth. In short, our understanding *is* our mode of 'being in the world.'"[16] This view of cognitive functioning and the relation of language to thought contrasts with what Lakoff and Johnson call an "objectivist" view: that meaning is representable solely in terms of propositional statements. The objectivist assumes that we have access to objective, mind-independent knowledge of the world. These conflicting points of view are in evidence when we consider how we understand change.

Change and Causation

Humans are part of an ever-changing world. Early in life each of us learns to function optimally in the world by responding appropriately to changes as they occur and by imposing changes on the environment to suit our needs and desires. From infancy on we build complex, largely unconscious cognitive understandings of what causes things to happen, of change, actions, states, and purposes. As with all of our human conceptual structure, these understandings are the product of our interactions with the world. Because those interactions are many and varied, we have multiple ideas based on our embodied experience of each concept (e.g., of causation).

When we lift a book from a table we have a direct experience of force against which we operate to lift the book. When we exert force against an open door, we cause it to close. Direct experiences of manipulating objects by exerting a force, the use of our bodily capacities to effect change, provide our most fundamental concepts of causation. Most typically, the application of force causes a change in location of the object acted upon, as when the book is lifted or the door is closed. These direct experiences form the basis of our metaphorical concepts of change and causation.

In one of the most fundamental metaphors for our understanding of events and causes, change is conceptualized as physical movement between one state and another. In this complex metaphor the source domain is motion in space. The metaphor rests on our varied and intimate knowledge of motion that derives from our actions on objects. The target domain is the domain of events as we perceive them. The "state" to which the metaphor applies may be the physical state of an observed entity or something more abstract, such as an emotional state. A state or set of conditions is metaphorically conceptualized as a bounded space, such as a container:

> The house is *in* good condition.
>
> She is *in* a terribly depressed state.
>
> At room temperature, water is *in* the liquid state.
>
> The molecule is *in* an excited state.
>
> A short pulse of radiation at the resonant frequency *puts* the electron *in* the excited state.

Change is expressed metaphorically as movement from one location (state) to another:

> His condition *has gone from* bad *to* worse.
>
> The weather *went from* sunny *to* stormy in just an hour.
>
> We had little success in preventing *progression* of the disease.
>
> The transition *from* the ground *to* the excited state is symmetry-forbidden.
>
> Amphiphilic block copolymers can self-assemble *into* ordered mesophases.

Because a change from one state to another is conceptualized as change in location, the rapidity with which a change occurs is expressed metaphorically as a rate of motion, that is, speed:

> Folding can be a *fast* or *slow* process, depending on the protein.
>
> A catalyst serves to *speed up* the rate of a chemical reaction.

Changes sometimes are seen as proceeding through distinct intermediate stages. Consistent with the concept of change as change in location, we speak of such overall processes as proceeding in "steps":

> She *took steps* to reorganize the office.
>
> The overall mechanism of the reaction is a *multi-step* process. The second *step* is rate-determining.
>
> We can visualize transcription as a *three-step* process.
>
> At this *stage* of the overall rearrangement process, the molecule is at the highest energy point.

An entailment of this metaphor for change, commonly found in scientific accounts, is that impediments to change are conceptualized as "barriers." Systems that do not change are conceptualized as "isolated" from factors that could produce change. Here are some examples taken from recent scientific literature (italics mine):

> The clusters are formed in the high-density, relatively hot region of the expansion, where there is still sufficient energy to *surmount* any *barriers* on the potential energy surface to reach the global minimum.[17]
>
> The *generation* of T cells capable of transferring diabetes *is blocked* in the absence of GAD expression in the beta cells.[18]
>
> In addition, the qubits must be sufficiently *isolated* from the outside world so that interaction with such reservoirs does not disturb. . . .[19]

We regularly use words such as "speed," "rate," "fast," "slow," and "multistep" in their conventional, literal senses to talk about motion in space. Similarly, the words "barrier," "block," and "isolated" are used literally to denote impediments to movement or change. Yet the same words are used in the preceding examples to talk about changes in which physical movement, if any, is incidental. Notice the systematic character of these appearances. Entire families of words that apply to motion in space in the literal, macroscopic world in which we go about our lives are applied to another domain. This systematic character, and coherence in mappings such as those just cited, is evidence for a general mapping from the physical domain of motion in space to the more abstract idea of change in state.

We also find that many words from the domain of social interactions, such as "mediate," "facilitate," "co-opt," and "regulate" are used to characterize features of processes at the microscopic level, as in this example from the literature (italics mine): "This is attributable to the *co-opting* of existing brain blood vessels by the implanted tumor cells."[20] The role of an agent involved in a change is conceptualized in terms of a familiar role played by humans in the social domain.

The conceptual metaphors that underlie all the examples given are the products of unconscious cognition, grounded in our everyday experiences and observations. The systematic character of multiple uses of words, as in the examples just cited, tells us that these are not just cases of using words arbitrarily for multiple meanings. Rather, they result from mapping of the concrete domain of force producing movement onto more abstract domains of change.

✳ ✳ ✳

The foregoing metaphor may be the most commonly used metaphor for change, but it is not the only one. In an alternative way of conceptualizing change, the focus is on attributes.[21] The metaphor has these elements:

> Attributes are possessions.
>
> Changes are movements of possessions (acquisition or loss).
>
> Causation is transfer of possessions (giving or taking).
>
> Purposes are desired objects.

Here are some examples:

> He *lost* his patience.
>
> Somehow she *acquired* the courage to make changes in her life.
>
> Folding *gives* the protein a more compact structure.
>
> In a single mutation the organism *acquires* immunity.

The goal is to find a superconductor that *loses* resistance to the flow of electricity at a temperature above room temperature.

The two basic metaphors for change are related in this sense: In what we will call the location metaphor, the entity undergoing the change is conceptualized as moving from a location identified by one set of properties to another with different properties ("The oxide went from normal to superconducting at a pretty high temperature"). In the alternative metaphor, the entity undergoing the change is conceptualized as remaining in place and receiving or losing properties of interest ("The oxide gains entropy as its temperature increases"). The main point to note is that the two metaphorical forms are based on closely similar reasoning patterns. Both are based on the concept of motion in space. Both are consistent with the idea that our reasoning about the important concept of change is embodied.

Causation

Because we have needs and desires, and we act to satisfy them, the idea of causation as purposeful action is fundamental to our understanding of the world. In Western philosophy, causation is thought of mainly in terms of what Aristotle called "efficient causation": Change is attributed to the application of a force or the existence of a prior necessary condition. There is a literal connection between the causative agent and its effect. The connection often is not clearly evident from the observational data at hand. For example, does the daily consumption of yogurt lead to a longer life? Such questions can be addressed through mathematical analyses of what constitutes legitimate cause-effect relationships.[22] However, such analyses are not concerned with the ways in which the causal relationship is conceptualized.

In the theory of conceptual metaphor, much of our understanding of causation is seen to be metaphorical, not literal. Of course, there is direct, literal causation and change, as in "The golf ball hit the window and broke it." But when we are reasoning about change and causation in more abstract domains, we use conceptual metaphors.

As we have seen, two basic metaphors dominate in much of our reasoning about change. In the location metaphor, causation is conceptualized as forced movement. The verbs that are appropriate to this metaphor therefore relate to movement. Here are some examples from everyday usages:

The medication eventually *brought her out* of her coma.

His commencement speech *moved me* to tears.

Heating *brings* the liquid *to* a boil.

In these examples a change from one state to another is conceptualized as *movement*. In each case the cause of the movement, the change in state, is an entity—"medication," "commencement speech," or "heating"—that is not literally capable of effecting movement. The following examples, typical of scientific accounts, also describe change as movement. We can readily identify the causal agent and the verb that denotes movement:

> Radiation at the resonant frequency *puts* the electron *into* the excited state.

> Neuronal activity can *elevate* serotonin concentrations.

> A reduction in food supply could *produce a* major *shift* in marine populations of the deep ocean.

Where change is conceptualized as movement from one location to another, causal agents are many and varied. Change may be facilitated when barriers are reduced or impeded when barriers are formed. Once again, it must be said that we don't ordinarily think of the language used in these examples as metaphorical. Seen from the viewpoint of conceptual metaphor, however, they tell us important things about how we humans reason about change in the world.

Reasons and Purpose in Causation

Humans regularly draw up mental plans to carry out certain purposeful actions. When the plan is carried out, our actions often enough achieve our desired purposes. We therefore develop a conceptual understanding that causation is action taken to achieve a desired purpose. In any given instance the cause of the action is the reason why the action will achieve the intended purpose. In reasoning about change, people often resort to teleology, the idea that there are purposes underlying change. Here are some examples drawn from the scientific literature:

> Knowing how reach plans are represented in the brain can tell us much about the mechanisms and strategies the brain uses to generate reaches.[23]

> By forcing the buildup of a protein that prevents NFkB activation, PS-341 seems to starve tumors of their blood supply and growth stimuli, thereby promoting their self-destruction.[24]

> These microbes infect cells and enlist several of the components that cells normally use to extend lamellipodia to power the bacteria's own travels within the host's cytoplasm.[25]

> Relaxin has diverse actions in the reproductive tract and other tissues during pregnancy. These actions include promotion of growth and dilation of the cervix,

growth and quiescence of the uterus, growth and development of the mammary gland and nipple, and regulation of cardiovascular function.[26]

In these examples entities such as body organs, proteins, bacteria, and drugs are understood to be carrying out their characteristic functions as though they were purposeful, self-directed agents. Linguistic use of this kind is common in scientific accounts, although in some quarters it would not be considered good form in formal science communication. Whether frowned upon or not as a stylistic device, teleological metaphors are widely used. They stand as further examples of the many ways in which our conceptualization of the world is founded on metaphorical mappings from other domains of experience. In this case the mapping is from purposeful human actions to the envisioned causative actions of molecular or cellular components of living systems.

Implications of Conceptual Metaphor Theory

The theory of conceptual metaphor casts metaphor in a very different light than approaches based on grammatical or semantic analysis of figurative language. It suggests that metaphor plays an extensive role in the way we interpret individual experiences and relate one kind of experience to another. The metaphorical underpinnings of our conceptual systems are evidenced in our use of language, but according to conceptual metaphor theory, metaphor is much more than a matter of just language. Our experientially grounded metaphorical understanding of abstract concepts influences our thought patterns and actions as well as the ways in which we express ourselves.

The philosophical systems in vogue during much of the twentieth century banished metaphor to a realm outside cognitive significance, granting only that metaphorical constructs might have heuristic or pedagogical value for science. Kittay describes the situation:

> However, it was clear that science made use of "models." These models must be understood as extended metaphors—not literally true, but useful representations of the phenomena which often led to fruitful theoretic conceptions and new empirical discoveries. Examples such as the billiard-ball model of gases or the wave models of sound and light were cited as demonstrating the importance of models in the construction of scientific theories. The positivists' response was to say, in a fashion analogous to granting metaphors an emotive meaning distinct from a cognitive one, that models had a merely heuristic value for science—but then discoveries could be guided by almost anything: dreams, fortuitous findings, a random remark.[27]

Logical empiricism and, more recently, much of analytic philosophy have given way to empirical evidence from cognitive sciences that shows how humans interact with the world and how they interpret those interactions. The analysis of ordinary language and its uses and acknowledgment of the influences of the social milieu and of an important role for the scientist's intuition and tacit knowledge in scientific discovery have all played a role in ascribing to metaphor a significant place in cognition.

We might well ask whether conceptual metaphor theory carries any special implications for our understanding of how science is done. What light, if any, does it shed on the ways in which scientists model the physical world, design experiments, account for observations, and formulate and test theories? I believe that if we accept the major premises of conceptual metaphor theory, we are forced to recast our picture of how scientists work. The evidence summarized briefly in this chapter points to the following major conclusions:

1. Scientists understand the world largely in terms of metaphorical concepts.

2. In carrying out their activities, scientists use the same conceptual frameworks that they apply to other aspects of everyday life.

3. The most fundamental of those frameworks are based on embodied understandings of how the world works. They derive from the earliest and most pervasive interactions with physical surroundings and involve fundamental notions such as verticality, distance, front-back, and in-out.

4. Many conceptual frameworks used in reasoning about the physical world derive from experiential gestalts, ways of organizing experience into a structured form. These gestalts include those drawn from the scientist's social interactions with individuals, social groups, and society at large.

5. We humans give meaning to our perceptions of the world and interpret data from the world (including extensions of the senses in the form of scientific instruments) largely in terms of metaphorical understandings based on embodied, unconscious reasoning. This contrasts with a hardcore realist position, which would be something to the effect that there is a direct, literal mapping from terms we use to describe the world to things as they are in the world. As a crude illustration, consider the statement that Athens, Greece is hotter than London, England. This seems quite straightforward; the statement is true if Athens is at a given time hotter than London or false if it is colder. In this example, realist thinking would have it that the statement either does or does not correspond to things as they are in the world. But to *what* things as they are in the world? The statement depends for its mean-

ing on what we understand by "hotter." The very concept of hotter or colder, of temperature itself, is the product of human reasoning, grounded in embodied experience. It does not exist independently of human thought. If there were no humans around to decide what is meant by "hotter," there could be no independently existing truths grounded in the concept.

In this way of looking at things, truth is the product of human reasoning. It follows that science does not proceed by discovering preexisting truths about the world. Rather, it consists in observing the world and formulating truths about it. As will become evident in the chapters ahead, much of what we regard as scientific truth is metaphorical representation.

This does not mean that science is capable of yielding only subjective results of uncertain reliability. In development from conception on, all humans undergo largely the same kinds of directly emergent experiences. We therefore possess closely similar conceptual frameworks insofar as embodied understandings are concerned. Because this is so, we are able to communicate with one another about a host of matters and to convey thoughts of great sophistication and subtlety. In the same way, communication between scientists rests on a large body of shared, directly emergent experience with the world. However, although scientists ordinarily have essentially the same directly emergent physical experiences, they may have significantly different social developments and therefore may have different understandings of social values, forces, and interactions. These differences doubtless lead to differing conceptualizations of events in the physical world based on metaphors drawn from the social domain.

6. Given these claims, it follows that modes of reasoning and communicating in science are not fundamentally different from those used in other forms of intellectual endeavor. Scientists apply the same tools of embodied reasoning in carrying out their scientific work that they use in other dimensions of their lives. The systems studied by scientists can often be made simpler (e.g., through the control of variables in experimental work). It is often possible to achieve a high degree of consistency in observations, leading to agreements on standard values for quantities, such as the speed of light in a vacuum.[28] The scientist's ability to control the complexity of the observation system is the basis of the vaunted reproducibility of many scientific results. But independently of the issues of experimental control, accuracy, and precision, scientists' understandings of scientific results, expressed in hypotheses,

models, and theories, are thoroughly embedded in unconscious cognitive processes and conceptual metaphor. There is no characteristic scientific rationality that stands apart from, let alone is superior to, rational thought applied to other spheres of human experience and knowledge.

If this point of view is correct, metaphorical thought grounded in deeply ingrained physical and social experiences must play essential roles in science. In the chapters ahead we will examine a selection of important metaphors drawn from various areas of science. My purpose is to present descriptive accounts of the ways in which metaphorical reasoning has shaped many important theories and hypotheses. I have not attempted to present exhaustive accounts but instead have selected from the historical record episodes that illustrate the development of individual metaphors as experimental evidence accumulates or that illustrate the ways in which social factors help to determine metaphor choice or interpretation.

We begin by considering atoms. The idea of the atom as the fundamental unit of matter has had a very long run in Western science. Although today it is thought of very differently from its original conception as the fundamental, indivisible particle constituting all matter, the atom remains the fundamental chemical unit. After all these centuries of speculation, search, and scrutiny, what do we know about atoms that is literally true?

The ancient Ionian Greek cities of Ephesus and Miletus, near the Mediterranean coast in Turkey, are popular tourist attractions. Ephesus, especially, has been uncovered and restored sufficiently to provide a sense of the splendid, prosperous, and energetic place it must have been. The beautiful façade of the library stands as testament to its status during the fifth and sixth centuries B.C.E. as the most extensive collection of materials in the Western world after Alexandria.

Both Ephesus and Miletus are now situated several kilometers from the sea, but for most of the time these cities were in their ascendancy, from the sixth through the first century B.C.E., they were busy seaports and prosperous trading centers. Ships came and went from all parts of the Mediterranean, and commercial traffic moved in and out to civilizations and cultures to the west and south. The more affluent Ionians traveled widely. One of the most famous Miletan explorers, Hecataeus (born about 540 B.C.E.), traveled as far west as Gibraltar and explored the Dardanelles and the coast of the Black Sea. The ideas, attitudes, and technologies that these travelers brought to Ionian city-states made for a lively and stimulating society. It was in this environment that Western science had its beginnings.

The Ionians made a significant departure from previous thinking by attempting to account for the primary causes of what they saw as systematic, reproducible changes in the natural world. The story begins with Thales, a prominent merchant of Miletus, who made many important innovations in early ge-

4

THE CLASSICAL ATOM

ometry. Unfortunately, nothing that Thales wrote has survived. We know of him chiefly through Aristotle's descriptions of his contributions. According to Aristotle, it was Thales who first gave thought to the question that became of overriding importance for the Ionian philosophers: "Of what is the world made?" The Ionians were looking for unifying principles, a governing causation to explain the great diversity and transient character of the observable world. Thales conceived of a single primordial substance from which all else derives. He sought to identify a mobile, fluid essence that pervades all change, that is present in everything even as the appearances change. His answer? Water.

Thales's Miletan successors, Anaximander and Anaximenes, proposed differing numbers and identities for the primordial substances. Heraclitus (540–475 B.C.E.), a resident of Ephesus, saw fire as the central element in natural causation. He identified fire with logos, the universal harmony and intelligence governing all things. The Ionians used the names of natural materials, such as water, fire, air, and earth, to denote primordial substances, but they were not intended as literal usages. Rather, the names represented different essences that, by changing in themselves (e.g., by vaporizing or condensing) or by combining with other primordial substances, could account for all the diversity and change of the material world.

The direction undertaken by the early Ionians found its most enduring statement in the work of Empedocles (500–430 B.C.E.), in far-away Sicily. Empedocles proposed that the world is formed of four primordial substances of equal importance: fire, air, water, and earth. All material things are made up of these four substances in some proportions. As before, the four substances were not thought of as elements in the modern sense of the word. "Water" could apply to other liquid substances, "earth" to any of several solid materials, and "air" to any gas. Associated with the four primordial substances are four primary qualities: heat, cold, moisture, and dryness. Empedocles' ideas are encapsulated in the famous diagram shown in figure 4.1.

In Empedocles' conception, change consists of the mixing or separating of proportions of the four primordial substances. But there is more involved than merely bringing them together. Two opposing tendencies, "Love" and "Strife," govern all interactions of the elements. Mixing to form a new material occurs under the influence of the agency "Love." Substances that seem to repel one another rather than mix act under the influence of the agency "Strife," or "Hatred."

The metaphors that underlie views of the material world held by Thales and his successors were, at the more general level, of a single kind. These early philosophers and observers of nature conceptualized causation as a material thing. A conception of this kind is a natural outcome of our embodied experiences with

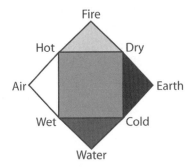

Figure 4.1. Empedocles' diagram relating the four "elements" and the characteristic properties of matter.

characteristic behaviors of things. Lakoff and Johnson refer to the folk theory of essences, that everything has an essence that makes it the kind of thing it is: "An essence is a collection of natural properties that inheres in whatever it is the essence of. Since natural properties are natural phenomena, natural properties (essences) can be seen as causes of the natural behavior of things. Thus we say things like, 'Oaks have leaves because they are trees.'"[1] Underlying this seemingly straightforward literal statement is the metaphorical understanding that a property of oaks, that they are trees, is the cause of their having leaves. For the Ionian philosophers, it is but a step to ask whether there is a more general, primordial causation underlying the multiplicity of events in nature. For Thales, causation resides in water, the primordial substance that inheres in all matter and underlies all change in matter. We can see how water, or some essence of water, might be chosen as the primordial substance. It is present in all living things. When living things die, they dry out. Water exists as solid, liquid, and vapor and can be seen to change between these forms of matter.

It is not clear how Thales envisioned the formation of the entire material world from water as the one primordial substance, but that is beside the point. The two most significant elements of his worldview are these:

A basic causal metaphor to account for events occurring in nature.

The insistence that there is regularity and systematicity in nature. Causation lies in nature itself, not in the whims of anthropomorphized gods and goddesses.

Later philosophers, by incorporating multiple primordial substances, more readily accounted for change as a coming together or separation. We can recognize elements of the location metaphor for change in the idea that bringing primordial substances together, or separating them, corresponds to a change in state. The causation metaphor that conceptualizes change as a gain or loss of properties underlies the idea that the loss or gain of a primordial substance causes a material to change from one form to another. Causation arises in either of the opposing notions of "Love" (attraction) or "Hatred" (repulsion). These anthropomorphic notions are recognizable as metaphorical mappings from the domain of human social interactions.

Enter the Atom

Empedocles viewed all existing substances of the material world as resulting from combinations of the four primordial substances. But the world is highly varied. How can there be so many different substances, with such differing properties? Empedocles hypothesized that any given substance is formed from some particular, fixed combination of the four primordial substances. Inasmuch as these four elements could be combined in an arbitrarily large number of different ratios, there is no problem in accounting for diversity in the observed world. But an important entailment is that any given material is composed of a characteristic and invariant mixture of some or all of the four primal ingredients. Thus, Empedocles' model holds within it the concept of the law of constant proportions, which played an important role 2,300 years later in Dalton's chemical theory of the atom. Empedocles did not propose the existence of atoms, but he proposed that each primordial substance is formed from invisibly small, homogeneous, and changeless particles.

Leucippus of Miletus is generally credited with originating the idea that matter is composed of atoms. Perhaps because none of Leucippus's writings survived, his pupil Democritus is more widely associated with the concept and was undoubtedly responsible for much of its development. Democritus proposed the existence of indivisible, invisibly small elementary corpuscles of matter, called atoms. Just as importantly, he proposed the existence of the void or empty space. The concept of space in which nothing exists was a new and controversial idea. It provided a conceptual basis for the idea of movement, which in turn provided for the possibility of change in the microscopic world that lay beneath the reach of the senses.

The atoms were thought to be infinite in number and to have existed eternally, in constant motion in an infinitely large space. They might differ from one another in shape, arrangement, and position, but the properties of any given atom are permanent. The motion of atoms was thought to be continuous and random, governed by mutual collisions. Aggregation to form the various substances of the observable world could occur if the atoms' shapes, sizes, and relative positions were appropriate. Such aggregations do not cause the atoms to lose their identity; they remain the same impenetrable units, and they continue to be separated by an absolute void. As the substance of which they are a constituent undergoes change, the atoms may be released to form other new substances. We observe the constant dance of apparent change, but the underlying reality is one of immutable atoms and void.

In searching in embodied experience for the origins of the metaphors that make up the Miletan atomists' view of nature, we must remember that they were not scientists in anything like the modern sense of the word. Their hypotheses were based not on experimental studies but on general conceptions formed through everyday experiences. Among these ideas was that change consisted in either fusion or pulling apart. Thus, movement was seen as essential to change, in keeping with the underlying metaphorical understanding of change as motion in space. The idea of atoms as being in constant motion might have arisen by analogy with the appearance of a swarm of gnats, or (as we shall see shortly) the movement of dust motes in a beam of sunlight. The notion of immutability was consistent with experience with forms of matter that seemed for all practical purposes to be immune to change, such as many minerals. Many items commonly encountered in the everyday world would have been formed to fit together with other pieces to form a whole object. It would be natural enough to conceptualize different atomic forms as having different shapes and sizes, fitting together with certain other atomic forms and not with others.

The concept of a void introduced a truly new primordial entity. It is by no means intuitively obvious, although once the concept of atom has been formed, the idea of empty space in which the atoms can move might follow naturally enough. Lacking such space, how could atoms be free to encounter one another? Lacking such encounters, how could change occur? Or so the reasoning might have gone.

A second new idea introduced by the atomists was that of the random motions of atoms. This represented a different idea from the notion of attraction or repulsion, a "Love" or "Strife" factor that determines the combinations of the primordial substances in Empedocles' model. In the atomists' model, change is determined to a large extent by chance. When atoms encounter one another, they may or may not fit together to form a substance. There is little of the anthropomorphic sense of love or strife in the atomists' conception of change.

A secondary consequence of the atomic view is that the four substances proposed by Empedocles are no longer primordial but are formed from the more basic primordial atoms. The four elements now become something akin to what we would think of in modern terms as chemical compounds.

In summary, these are the properties metaphorically attributed to atoms by Democritus:

       Invisibly small

       Hard, impenetrable

       Characteristic shapes and sizes

       Constant random motion in a void

       Combination to form the observed material world

The atomist view of matter found a full expression and significant extensions in the writings of Epicurus, who lived about a century later than Democritus. To the attributes just listed, Epicurus made a fateful addition, that of weight. He also proposed that all atoms move about at the same speed, regardless of their weight and volume. But a problem immediately arose. If the atoms have weight and motion, they should have rectilinear (straight-line) motion downward, as was believed to be natural to all bodies. In the observable world, objects of weight fall when raised above Earth's surface and released. The model thus seems to entail the unappealing prospect of all the atoms piling up at the "bottom," wherever that is. It hardly provides for the possibility of interactions leading to new substances. And what of the air, which floats around us?

To escape the consequences of his proposal that atoms have weight, Epicurus proposed a certain kind of motion, the clinamen, which allowed atoms to swerve in some way from their natural downward paths. They were in this way free to collide with one another; scattering from the collisions in turn would lead to random motion. The clinamen was a rather unsatisfactory patch to somehow preserve features of the model that were deemed essential, and it served as a focal point for Epicurus's critics.

The poem "De Rerum Natura" ("On the Nature of Things"), by the Roman poet Lucretius, who lived in the first century B.C.E., summarizes admirably the state of understanding of nature achieved by the Epicureans. Lucretius did not concern himself much with the ethical aspects of Epicureanism but instead produced a beautiful treatise on the physical theory in Epicurus's account. The poem is an impressive collection of hypotheses and theories of the physical world, a ringing renunciation of religious superstition and a denial of divine power. Lucretius believed in the power of embodied reasoning based on our sensory experiences of the world: "The senses that are common to all men indicate that matter exists. Unless confidence in sensation is first firmly established, there will be nothing by reference to which our reason can reach any conclusion regarding things not perceived by the senses."[2]

The following passage suggests how the hypotheses of atoms as invisibly small particles in constant motion might have arisen from direct observation:

> Whenever the rays of the sun stream in and pour light through the dark places of the house, look and you will see many minute particles darting about in many directions through the empty air in the light of the rays. . . . You would do well to observe these motes which you see dancing in the sunbeams: this dancing indicates that beneath it there are hidden motions of matter which are invisible. You will see that many motes, struck by unseen blows, change their courses and are forced to move now this way and that, on all sides and in every direction. Truly this change in the direction of all motes is caused by the atoms. The atoms first move by themselves; then the complexes, which are small gatherings of atoms and are, so to speak, next to the atoms in force, are moved when struck by the invisible blows of the atoms, and these in turn set in motion particles that are a little larger.[3]

Lucretius's explanation of the movements of dust motes in sunlight is incorrect; it is typically small currents of air that give rise to the motions. Nonetheless, his reasoning from observations in the macroscopic world to a detailed explanation in terms of entities in the microscopic world is remarkable. The reliance on experiences with forces, impacts, and directions in the everyday world to understand the processes at work in the microscopic domain provides an excellent example of embodied reasoning.

Elsewhere in Lucretius's poem there are speculations that relate the observed properties of substances to characteristics of the atoms. Solids, which are hard and compact bodies, were thought to have "crooked" atoms that could become more intimately linked and intertwined. Liquids were thought to be composed of smooth, spherical atoms. Gaseous dispersions such as smoke, clouds, and flames were visualized as having pointed extremities that could cause them to penetrate and fit into other materials. Taste received an especially elaborate treatment. Pleasant, sweet tastes were thought characteristic of atoms with smooth surfaces, whereas bitter tastes were associated with crooked atoms with sharp edges. Even color was associated with the shapes of the atoms. We see here an evident metaphorical mapping from human sensory experiences in the everyday world to the presumed properties of atoms in the microscopic, invisible world.

It is noteworthy that in the Ionian model the observed properties of matter were associated with individual atoms, not with the clusters that formed when atoms came together. The notion of associating properties with aggregates of atoms came later but still well in advance of a modern theory of materials. Again, we must keep in mind the important fact that these speculations did not inform a program of experiment. The idea of manipulating nature in some way to test a hypothesis had not yet arisen.

The Atom's Long Slumber

In all of Plato's writings there is no mention of Epicurus or any of his work. Plato apparently detested the Epicurean philosophy and model of the world. His own conception of the fundamental nature of matter is of little interest other than to note that he conceived of Empedocles's four elements as built on distinct geometric forms. The Epicurean model did not fare any better with Aristotle, who was decidedly anti-atomist. His philosophical outlook was vastly different from that of the Epicureans. He believed that matter was indefinitely divisible. Aristotle saw events in the world as the product of rational intention. Although his notion of a material cause corresponded roughly to the Ionian concept of causation, his conception of causation was much more complex. Aristotle came down strongly on the side of Empedocles's model of the four primordial substances. With his endorsement and elaborations, that model became the standard interpretation for many centuries.

Plato and Aristotle were the philosophers of antiquity whose views were widely known and accepted in the Christian era. This happened in part because their work survived largely intact, but there were other considerations. The early church leaders, such as St. Augustine (354–436 C.E.) and St. Basil (330–79 C.E.), condemned many principal elements of Epicurean philosophy, such as the idea of a boundless universe, the concept of the void, and the role of chance. The Epicurean rejections of a divine power that governed the universe and of a life of the soul after death were also contrary to the teachings of the church.

The atomist model of the material world lay in relative obscurity, at least throughout Christian Europe, for many centuries. The occasional writer who revived it, even after attempting to reconcile atomism with church doctrine, often paid dearly for his efforts. It became increasingly evident from the fourteenth century onward, however, that Aristotelian physics did not provide an adequate explanation of how the world worked.[4] By the time of Galileo, in the early seventeenth century, it had become at least quasilegitimate to speculate about the microscopic nature of matter in other than Aristotelian terms. Galileo openly endorsed the atomist theories of Democritus and Epicurus, for example, by expressing the conviction in 1638 that a vacuum could exist.

At about the same time, the atomist cause received a strong impetus from the activities of Pierre Gassendi, a prominent French cleric and philosopher. Gassendi rejected the Aristotelian account of nature and undertook to revive Epicureanism. To do so, he attempted to reconcile mechanistic atomism with Christian doctrines regarding the soul, immortality, and free will. Gassendi also saw to the printing of Lucretius's poem, "De Rerum Natura," in 1649. The atom's long slumber was drawing to a close.

The Era of Experiment

In the mid-seventeenth century, experiment began to play an important role in the making of new knowledge. As described by Steven Shapin, "This stress on artificially contrived experiments is nowhere more apparent than in research programs associated with the Royal Society of London (founded in 1660) and especially with its most influential fellow, Robert Boyle. The air pump invented for Boyle by his assistant Robert Hooke in the late 1650s swiftly became emblematic of what it was to do experimental natural philosophy. It was the Scientific Revolution's greatest fact-making machine."[5]

Boyle's experimental apparatus was used to settle decisively a contentious matter that had arisen several years earlier, as a result of experiments of Evangelista Torricelli. This Italian mathematician, an admirer of Galileo, had performed a very simple experiment by filling with mercury a long glass tube sealed at one end and then inverting the tube in a pool of mercury. The experiment is shown schematically in figure 4.2. (Torricelli's apparatus is called a barometer.) Torricelli found that the mercury in the tube became steady at a level about 76 cm, or 29 inches, above the mercury in the pool. He claimed that the column resulted from the weight of the atmosphere pressing down on the surface of the mercury outside, forcing the mercury up the tube. Because there is no air in the tube, there is no corresponding force pressing down on the mercury in-

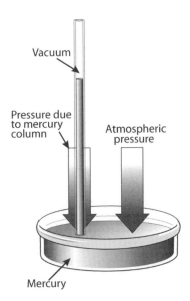

Figure 4.2. Torricelli's barometer.

side. The experiment thus demonstrated, according to Torricelli, that we live at the bottom of an ocean of air; the atmosphere has weight.

Torricelli's interpretation contradicted at least two well-entrenched beliefs. Aristotelian doctrine had it that air in its proper place, such as the atmosphere, effectively had no weight. Second, there was the question of what could be in the space above the mercury in the column. Torricelli answered that there is nothing there; it is a vacuum, in contradiction of the conventional wisdom that nature abhors a vacuum. Descartes had said that nothing can be filled with Nothing.

There is a story that the famous French philosopher Blaise Pascal arranged to have the Torricelli apparatus carried to the top of Puy de Dome, a volcanic peak in central France, to have the height of the mercury column compared with that for a duplicate apparatus at the foot of the mountain. Its height was reported to be about three inches lower at the higher elevation. The result is predicted by Torricelli's model; at the higher elevation, some of the atmosphere is below the apparatus, and therefore there is less of it to press down on the mercury surface.

Boyle delivered what should have been the coup de grace to any lingering opposition to Torricelli's model. He placed the apparatus in a large glass container and then pumped out the atmosphere. As the container was being pumped out, the height of the mercury in the tube steadily fell and eventually was nearly at the level of the mercury in the dish. When air was readmitted to the container, the mercury column again rose. Boyle's experiment did more than simply extend Pascal's experiment. There could be no question of the weight of the atmosphere pressing on the mercury surface while it was in the glass container. After the container had been effectively pumped out, and air readmitted, only the air in the glass container, and not a column of it extending to the top of the atmosphere, could be responsible for the difference in mercury levels. Boyle accounted for the observed effects by referring to the "spring of the air": "Imagine the air to be such a heap of little bodies lying one upon another, as may be resembled to a fleece of wool. For this . . . consists of many slender and flexible hairs, each of which may indeed, like a little spring, be still endeavoring to stretch itself out again."[6] Notice that this metaphor for gases lacks an atomic character. Furthermore, there is little sense of motion on the part of the constituent particles or of any relationship between temperature and gas behavior.

The opposition was not immediately quelled by Boyle's results. The philosopher Thomas Hobbes remained convinced of the philosophical impossibility of a vacuum. Boyle was not much given to philosophical speculations. He replied to his critics in part by carrying out a series of experiments in which he measured the volume of a given quantity of gas as a function of the pressure exerted on the gas. He found that when the gas was compressed, the pressure increased and the volume of gas decreased. His quantitative measurements led to what is known as Boyle's law: At constant temperature, the volume of a given sample of gas varies inversely with pressure. This means that the pressure times the volume is a constant for a given gas sample at constant temperature. Boyle's law can be written in the form of an equation, in which P represents pressure, V is the volume, and C is a constant that depends on the amount of gas and the temperature:

$$PV = C \qquad\qquad\qquad\qquad (eq. 4.1)$$

This simple equation represents the first example of a mathematical model based on data from deliberately planned experiments.

Boyle declined to speculate on the origins of the observed properties of gases. He wrote that his intent in the experiments was not to theorize about the causes of gas properties but only to demonstrate what those properties are.

Theory and Experiment

The Swiss mathematician Daniel Bernoulli (1700–1782) applied his analytical skills to problems across a wide range of scientific disciplines. He made major contributions to the study of fluid flow. Despite his eminence, one of his most important contributions lay neglected for more than a century. Bernoulli pictured gases as consisting of enormous numbers of particles in fast, independent motion. The collisions of the particles with one another and with the walls of the vessel containing the gas were assumed to be completely elastic. That is, there was no tendency for the particles to stick to one another or to the walls. He postulated that the speeds of the particles were related to temperature; higher temperature meant higher speeds.[7]

This model for the gas entails that gas pressure is the result of the force on the vessel wall exerted by the gas particles colliding with the walls. The model led Bernoulli to a mathematical treatment that gave an account of gas behavior in agreement with experiment. For example, it led to Boyle's law. Furthermore, it accounted for why the pressure in a vessel increases upon heating: As temperature increases, the speeds of the gas particles increase. They therefore collide more frequently with the wall and exert greater force on the vessel wall when they do so.

Bernoulli's work is one of the earliest examples of reasoning from a model, a set of related metaphors, to a mathematical expression that expresses the content of those metaphors. Here are the major metaphorical elements of the model:

Atoms are in constant, chaotic motion.

Collisions with other atoms and with the vessel walls are elastic (atoms are like billiard balls).

The speeds of atoms increase with increasing temperature.

The average kinetic energy (i.e., energy of motion) of a gas atom depends only on the gas temperature, not on the mass of the atom.

Pressure results from collisions of gas atoms with the vessel walls.

The assumption that the average energy of the gas atoms is related to temperature was a striking innovation, at odds with prevailing ideas of the nature of heat.

Let's pause briefly to consider whether Bernoulli's model, contained in these listed elements, is indeed metaphorical. Someone might look at this with contemporary eyes and say, "Well, whether Bernoulli knew it or not, we know today that the gas literally consists of atoms moving about chaotically, and so on. This is now a literal model!" Well, we don't see atoms moving about in a gas, even today, but aside from that, consider the nature of the mapping involved. In the observation domain we have data giving the pressure, temperature, and volume of a gas as these related variables are altered experimentally. The model is an attempt to account for these observations. It is not a literal analogical mapping onto the observation domain but rather a thoroughly metaphorical representation of the gas in terms entirely different from the terms of the observation domain. In Bernoulli's hands, the model becomes a mathematical description that predicts the properties of gases that can be compared with the observations. Yes, this mathematical description contains the same quantities that make up the observation domain: pressure, volume, and temperature. But the relationship between these quantities in Bernoulli's equation is determined by the underlying model, with the elements listed earlier. Because the model is a metaphorical description, the mathematical description of gas properties, which refers to the model, is inherently metaphorical as well.

Many elements of this metaphor may seem similar to those advanced by the Greek atomists, but there are important differences. Bernoulli's model is not about atoms in general; it seeks to explain only the observed properties of gases. Therefore, it is not a model that relates atomic properties to taste, smell, and so on. Not even the sizes of the atoms are considered, other than the assumption that they are very small in relation to the distances between them. Thus, Bernoulli's model shares a property we saw earlier to be true of metaphors in general: They highlight certain features of the target domain and hide others.

Bernoulli's theory of gas behavior was an intellectual success by any standards of reckoning. The mathematical model formulated from a handful of basic metaphors accounted beautifully for the known properties of gases and was capable of predicting others. Yet it received very little attention and remained essentially unknown for more than a century. The ostensible reason for its lack of popular success is that it conflicted with two other well-known theories. The most obvious conflict was with Isaac Newton's theory of gases.

Newton accepted the atomic hypothesis, or at least some important components of it. Here are his words from *Optiks* (1719):

> All these things being considered, it seems probable to me that God in the Beginning form'd matter in solid, massy, hard, impenetrable, moveable Particles of such Sizes and Figures, and with such other Properties, and in such Proportion to Space, as most conduced to the End for which he form'd them; and that these primitive Particles being Solids, are incomparably harder than any porous Bodies compounded of them; even so very hard, as never to wear or break in pieces; no ordinary Power being able to divide what God himself made one in the Creation.[8]

Newton thus accepted the idea of atoms as the building blocks of the observable world while rejecting many of the elements of the Greek atomist worldview that conflicted with his religious beliefs.

On the basis of his acceptance of atoms, Newton formulated a model for gases that accounted for Boyle's results: He assumed that gas atoms repel one another, with a force of repulsion inversely proportional to the distance separating them. That is, the closer two gas atoms approach one another, the greater the repulsion between them. A gas in Newton's model consisted of a static assembly of atoms in the space occupied by the gas, each atom positioned so as to minimize its repulsions with all the other atoms present. Gas pressure arises because of the forces pushing the gas atoms out and away from one another. The more the gas is compressed, the greater the repulsive forces between atoms and the greater the observed gas pressure. Although Newton's model accounted for Boyle's results, it did not explain a variety of results that began to emerge as additional experiments were performed on gases. For example, it did not explain the observation that gas pressure increases as a gas is heated at constant volume. Yet Newton's reputation was so great that any theory he advanced was regarded as virtually unassailable.

Bernoulli's model also conflicted with the popularly held theory of heat. Joseph Black, a British scientist, had developed the caloric theory of heat, which proposed that heat is an invisible substance, called caloric, that permeates all matter. In this model, exchanges of heat between two bodies involved the flow of caloric from the hotter to the colder body. The idea that heat was in some way related to motions of atoms was not a familiar and accepted view.

More than a hundred years after Bernoulli had published his work, John

James Waterston, an English civil engineer and amateur physicist, independently developed the gas kinetic theory even further. He derived the ideal gas law, which has the form

$$PV = CT \qquad\qquad\qquad (\text{eq. 4.2})$$

Here P, V, and C have the same meanings as in Boyle's law, equation 4.1. The additional factor is temperature, T. If temperature is constant, it becomes part of the constant C. Then we simply have Boyle's law. Equation 4.2 is more general in that it also predicts what happens if temperature is changed. For example, if temperature increases, the product PV must increase. Waterston's ideal gas equation was the product of a sophisticated model for gases, but in 1845 he could not even get his work published by the scientific community. Part of the reason was snobbery: He was not seen as having the right credentials.

Experiments under way in this period were setting the stage for acceptance of a more complete theory of gases that incorporated Waterston's and Bernoulli's conceptions. James Prescott Joule, a British physicist, showed in a series of experiments that heat is a form of energy and that heat and energy are, in part at least, interconvertible. The caloric theory of heat was disproved conclusively. Second, he performed an experiment that demonstrated that Newton's theory of gases was incorrect. Joule caused a gas to expand into a vacuum. The temperature change observed was negligible, whereas Newton's theory required that there be a large change. With the two major conflicting theories out of the running, the way was paved for a complete theory of gases. James Clerk Maxwell and Ludwig Boltzmann, two of the most famous names in nineteenth-century physics, brought the theory to a high state of mathematical development in the 1860s.

We can trace through the succession of gas theories an increasing complexity and sophistication of the underlying metaphors. The model formulated to account for the pressure-volume behavior of gases evolved into a more general kinetic theory in which the motions of the gas particles are modeled in a more detailed fashion. The kinetic theory accounted for the behavior of gases but also served as a model for the relationship between temperature and energy. In the Maxwell-Boltzmann model, the gas atoms were assumed to possess average kinetic energies (i.e., energies of motion) directly proportional to something called absolute temperature. As absolute temperature increases, the average kinetic energy increases in proportion. Conversely, as temperature decreases, the average kinetic energy decreases proportionally. We can imagine reducing the temperature to the point where the kinetic energies are zero. That point is called absolute zero, the lowest possible temperature.[9]

Experiments eventually showed that real gases don't behave like the ideal gas of the gas kinetic theory, one more example of the dictum that models are never complete descriptions of reality. One class of experiments involved studies of diffusion. We are all aware that when an odor is introduced at one end of a room, as when a person wearing cologne enters and stops at the doorway, it is some time before the fragrance is detected at the opposite end of the room. Based on kinetic theory calculations, the atoms of a gas must be moving at about twice the speed of a modern jet aircraft. Why does it take so long for the fragrance to traverse the room? A possible answer is that the atoms (molecules) of the gas keep bumping into one another, like passengers in a busy subway station. They hardly get moving before they have a collision that deflects them from their path. By modifying the model for gases to take account of the fact that atoms have finite size, it was possible to account for diffusion behavior. Another correction to the simpler model was made to include the assumption that collisions between atoms may not be entirely elastic; the atoms may have some attractions for one another.

These additions were prompted by experimental results that showed deviations from the predictions of the simpler model. They followed in a straightforward way from everyday experiences with objects that have size and thus can get in one another's way and with things that may have some tendency to stick together. Translated into mathematical language, they provided expressions that account with high accuracy for the observed properties of gases.

Whereas the kinetic theory represented sophisticated ideas about the relationships between temperature and kinetic energy and about the ways in which the speeds of the atoms are distributed in the gas, the underlying metaphor for the atoms themselves had undergone no fundamental change. Although endowed now with a finite size and with an ability in some cases to exert weak attractive forces on one another, they remained featureless little things, moving about chaotically, colliding with the walls and with each other. But just how big are these atoms? What do they weigh? How many of them are there in a given sample of gas? Using the Maxwell-Boltzmann kinetic theory of gases, along with experimental data on gas properties, Joseph Loschmidt in 1865 tried to answer these questions. He knew that when a typical gas at atmospheric pressure is condensed at low temperature to form a liquid, the volume of the liquid is only about one thousandth that of the gas. So it seems that gas atoms are quite well separated in the gas. Furthermore, from measurements of diffusion of gas molecules (remember the cologne drifting across the room?), it was possible to calculate how frequently gas particles must collide with one another. Putting these considerations together, Loschmidt estimated that atoms are about one one-hundred-millionth of a centimeter in diameter. That is, one hundred million atoms aligned in a row would make a line about the diameter of a penny in length. He also estimated that a liter (slightly more than a quart) of an atomic gas such as helium or argon must contain about 24×10^{21} atoms. This is an unimaginably large number, on the order of a billion times larger than the U.S. federal debt in dollars. A way to grasp its magnitude is to imagine that the atoms in a liter of argon gas were condensed to a liquid and then each atom expanded to the size of a ping-pong ball. The argon would cover the territorial United States to a depth of several hundred miles!

The Chemical Atom

Our story of the atom has carried us well past the mid-nineteenth century. But it has been an incomplete story, focused solely on the metaphors for the physical properties of gases such as pressure, volume, temperature, and diffusion. We could call the atom we have seen so far the physical atom. To follow the development of an equally important concept of the atom, we must retrace our steps.

In the course of the seventeenth century the Epicurean model of atomism achieved extensive acceptance. But the physical atom envisioned by Bernoulli and Newton was in no way related to the old doctrine of primordial substances. In moving beyond this primitive view of matter, the chemists of the early eighteenth century began to study matter in a more quantitative way. The idea of a pure substance took hold: something that was composed of only one kind of material. Chemical changes of those pure substances, through heating alone or by reactions with other substances, were noted. The conservation of matter in chemical changes was established through quantitative experiments. The idea that matter is neither created nor destroyed during a chemical change is familiar to us today, but it was by no means taken for granted in the late eighteenth century. For example, many still believed that with prolonged boiling water was converted into earth. The great French chemist Antoine Lavoisier boiled water under carefully controlled conditions, collecting all the condensed water from the process, to show that there was no change in the weight of water. Lavoisier is most famous for discovering oxygen, which really amounted to showing that all forms of combustion known to that time involved combination of oxygen with the substance burned.

If there are many pure substances, it could hardly be the case that the great variety of them could be accounted for in terms of the outmoded notion of four primordial substances in varying proportions. Robert Boyle and then Lavoisier severed the link with the past once and for all by establishing the concept of a chemical element. Boyle defined an element operationally, as a substance that could not be further decomposed. He recognized gases such as oxygen, nitrogen, and hydrogen as elements. Among solid substances he correctly recognized many metals as elements: gold, silver, iron, tin, and so on. He also got quite a few things wrong (materials that seemed to him not to be further divisible that proved later to be so).

Notice that this operational definition of an element does not presuppose anything about the atomic nature of matter, even though Boyle, for example, adhered to a belief in atoms. But new results based on careful chemical work began to raise interesting questions. Joseph Louis Proust (1755–1826) noted that chemical compounds were formed with constant compositions. That is, whether found in nature or prepared in the laboratory, a particular substance had the same relative amounts of the constituent elements. His observation, which became known as the Law of Definite Proportions, was a marked departure from the lax constraints of older theories of the chemical nature of matter. Then a few very important chemical studies of gases forced further revisions in outlook.

A French chemist, Joseph Louis Gay-Lussac (1778–1850), showed that when gases react with one another, the ratio of the volumes of gases that combine generally is a small whole number. For example, when hydrogen gas is reacted with oxygen gas to form water vapor, two volumes of hydrogen react with one volume of oxygen, and two volumes of water vapor are formed.

John Dalton (1766–1844), one of the more extraordinary men in the history of science,[10] put these and other facts of his own making together. In 1808 he began publishing his *New System of Chemical Philosophy*. In the opening volume he proposed that elements were substances in which all the atoms are the same. When chemical reactions occur between elements, the atoms combine with one another to form substances that contain the elements in fixed ratios. For example, when hydrogen gas reacts with oxygen gas, water is formed. The most elementary particles of water, a compound substance, contain two hydrogen atoms and one oxygen atom.

Dalton made a crucial, erroneous assumption in working out the weight relationships between reacting gases. He assumed that all gases were monatomic; that is, that hydrogen gas is composed of single atoms of hydrogen, oxygen gas of single atoms of oxygen, and so on. In Dalton's view equal volumes of gases at the same temperature and pressure contained equal numbers of atoms. Several years later, the Italian Amadeus Avogadro proposed correctly that gas samples under the same conditions of temperature, pressure, and volume contain equal numbers of molecules. In hydrogen, oxygen, and nitrogen gases, the elements are present as diatomic molecules, which we write as H_2, O_2, and N_2. Dalton's mistaken guess caused rampant confusion for a half-century, but by about 1860 matters had become pretty clear. The chemical atomic theory had acquired a certain legitimacy, although it was by no means universally accepted.

Here are the elements of Dalton's atomic theory:

Each element is composed of extremely small particles that represent the ultimate level of divisibility of an element.

All atoms of a given element are identical. The atoms of different elements are different, including different masses.

Atoms are neither created nor destroyed in chemical reactions.

Compounds are formed when atoms of more than one element combine; a given compound always has the same relative number and kind of atoms.

We notice immediately that, except for the first, these metaphors overlap very little with the earlier lists drawn from experiences with the physical properties of atoms. Theories, and the metaphors from which they are constituted, are limited to accounting for a particular body of observations. They need not, and usually do not, draw upon or inform models that purport to account for significantly different evidences. For this reason, there could be a conflict between metaphorical frameworks based on different sets of observations of a given system.

The first of Dalton's metaphors arises naturally from embodied experiences, as noted earlier. The idea that the smallest divisible part of a substance is indeed small is supported by experiments that go well beyond the imaginings of the Greeks; it can be linked to the invisible particles that make up a pure gaseous substance, such as oxygen or hydrogen. Dalton's second hypothesis was not established by experiments performed in his time, nor could it be tested for more than a century. What Dalton asserted is mainly that a pure substance does not consist of atoms that vary greatly in weight. He thought that if there were large variations in weights of the atoms of a given element, they would tend to react selectively with another substance or to separate out in some way, in conflict with the Law of Definite Proportions.

The third element of Dalton's theory is derived from the Greek conception of the atom as continuing unaltered through the flux of apparent changes in matter. For Dalton, however, the immutability of atoms had a specifically chemical meaning: When an element reacts with another element to form a substance, and the substance is then decomposed, the original elements can be recovered.

The evidences that gave rise to the Law of Definite Proportions supported Dalton's fourth proposal. What it means is that when elements combine chemically, they do so according to rules that constrain the manner of combination. For example, carbon and oxygen react to form two common substances, carbon monoxide and carbon dioxide. The molecules of carbon monoxide contain one atom each of carbon and oxygen. The molecules of carbon dioxide contain one carbon and two oxygen atoms. There are no stable compounds of carbon and oxygen containing, say, six carbon atoms and eight oxygen atoms, or other complex ratios.

Certain elements of Dalton's theory proved to be controversial. In Dalton's time and for many years thereafter, it was not universally accepted that elements combine in constant proportions and generally simple ratios of atoms in forming compounds. The data were inconclusive, and it was not evident why atoms should be so particular in their combining capacity. But even aside from these problems, which could be related to technique, it was by no means clear what the term "atom" could be referring to. Humphrey Davy, a distinguished chemist of the time, insisted that "atom" could not be taken literally to mean small indivisible particles of matter but only to a unit of chemical reaction. That is, there was not seen to be any necessary connection between the chemical entity called "atom" and the physical atom of the kinetic molecular theory.

Anti-Atomists

At this point in our story we have two quite different kinds of metaphors for atoms, with not much in common between them. Both metaphors served well in guiding research in both physics and chemistry. Nevertheless, during the latter third of the nineteenth century, there arose a strong school of scientific thought that rejected the use of metaphors such as the atomic hypothesis in physics and chemistry. These scientists believed that the goal of science should be to achieve an "economy of thought" in descriptions of natural phenomena. In the view of this school of scientific philosophy, which derived at least some of its stimulus from Auguste Comte's philosophy of positivism, the use of hypothetical concepts such as atoms was to be avoided. The debates between advocates of the atomic hypothesis, such as Boltzmann, and its opponents, such as Pierre Duhem and Ernst Mach, were vigorous and highly visible. The extended disputes cast a pall over Boltzmann's kinetic theory of gases. They caused him deep depression and are thought by some to have contributed to his suicide in 1906.[11]

Ironically, just at this time there was new, strong evidence for atoms.[12] Albert Einstein was sympathetic to the concept of atoms and gave thought to how more conclusive evidence of their existence might be adduced. It occurred to him that if atoms (and molecules, formed by combinations of atoms) were in constant, chaotic motion, there might be evidence of their existence in the erratic movements of small particles drifting in solution and buffeted by the continual collisions with the molecules of the solvent. Einstein related later that he was initially unaware of the early-nineteenth-century work of the Scottish botanist Robert Brown, who had been the first to study such fluctuations. The effect was known, appropriately enough, as Brownian motion. Einstein worked out a quantitative theory of Brownian motion that yielded a probability distribution of colloidal particles in suspension. For an individual suspended particle in solution, the random motions could be accounted for in terms of the mass of the particle, the viscosity of the fluid, and the temperature.

The French scientist Jean Perrin used the newly developed ultramicroscope to observe the motions of large molecules suspended in a fluid, with a degree of precision and accuracy never before achieved. He showed that his observations were completely consistent with Einstein's theory of Brownian motion. He was able to estimate the sizes of atoms and molecules and their numbers in a given volume. Perrin's experiments, well designed and beautifully executed, made a very strong case for the discontinuous, atomic and molecular scale of matter. But he did not assume that the experimental facts and their interpretations would speak for themselves. He worked tirelessly to disseminate his results and to recruit other scientists to his point of view. The first Solvay Conference of Physics, in 1911, brought together the leading physicists from all of Europe. Perrin detailed his own investigations and summarized other evidences of the existence of atoms. His presentation was entirely convincing; no significant opposition to his conclusions was voiced, even though some of those present had been avid anti-atomists. In 1913 he published his book *Les Atomes*, which summarized all the evidences for the existence of matter at the atomic and molecular scale. By the standards of science books, it was a great success. Perrin's experimental results, coupled to his advocacy for his hypotheses, effectively ended skepticism about the existence of atoms. Perrin's work earned him the Nobel prize in physics in 1926.

Two decades before that time had arrived, still in the heat of battle with the anti-atomists, Boltzmann had this to say in a letter in the journal *Nature* in 1895:

> Every hypothesis must derive indubitable results from mechanically well-defined assumptions by mathematically correct methods. If the results agree with a large number of facts, we must be content, even if the true nature of facts is not revealed in every respect. No one hypothesis has hitherto attained this last end, the Theory of Gases not excepted. But this theory agrees in so many respects with the facts, that we can hardly doubt that in gases certain entities, the number and size of which can roughly be determined, fly about pell-mell. Can it be seriously expected that they will behave exactly as aggregates of Newtonian centres of force, or as the rigid bodies of our Mechanics? And how awkward is the human mind in divining the nature of things, when forsaken by the analogy of what we see and touch directly?[13]

Boltzmann's statement touches upon most of what this chapter has been about. Models are extended metaphors. Elements of the model map onto selected aspects of the observation domain. They are never complete descriptions of reality. In practice, no one model is sufficient to capture all the various observations that might be made. Thus, for example, we use distinctly different atomic metaphors to account for the physical and chemical attributes of matter.

A given model may itself suggest the ways in which discrepancies between the predictions of the model and observations might be resolved. Thus, the nonideal behavior of real gases was accommodated quite naturally by adding size and weak attractive forces to the simpler model of an ideal gas.

Models form the basis of theories. The model of a gas as spherical particles in chaotic, random motion gives rise to the kinetic molecular theory. The elements of a theory refer to specific entities in some model. Boltzmann reminds us that creative reasoning about the properties of the natural world draws on our embodied experiences. His words are particularly significant, coming as they do from a scientist whose *Lectures on Gas Theory,* a densely mathematical and difficult book, is one of the classics of the physics literature.

The story of metaphors for atoms is by no means at an end. As the nineteenth century drew to a close, a huge revolution in conceptions of matter at the microscopic level was getting under way. Just as it began to seem that atoms did indeed exist, and we could have some idea of what they were like, the Ionian era came unceremoniously to an end. The atom, it turned out, is not indivisible after all.

5

THE MODERN ATOM

In this chapter we will trace the evolution of the concept of "atom" through one of the most revolutionary periods in the history of science.[1] The old metaphors of billiard ball and immutable entity are replaced for some purposes by a succession of models in which the atom has structure, and there are increasingly quantitative relationships of model to observation. The history of the modern atom provides additional examples of metaphors that arise from mappings of embodied experience onto elements of the observation domain. I will argue that our various conceptions of the atom—for we can think of the atom in differing ways, depending on the observations being referenced—remain metaphorical in nature, even as they become increasingly mathematical. The story begins with the discovery of the electron, the first known subatomic particle. The existence of the electron meant that the atom is not the immutable, indivisible entity envisioned since the time of the Greeks; a new model would be needed.

The Electron

In 1897 the British scientist J. J. Thomson read a paper before the Royal Institution of Great Britain on his latest studies of the nature of cathode rays. He claimed to have discovered a negatively charged subatomic particle that is present in all matter. It has been said that many in the audience thought that Thomson was pulling their leg, that he couldn't possibly be serious about his sweeping proposal. But he was serious; in 1997 the scientific world celebrated the centennial anniversary of what is generally regarded as Thomson's discovery of the electron.

Thomson's work came after decades of study of cathode rays by scientists in many countries. The research in physics leading up to and including his contributions provides particularly interesting illustrations of the processes through which advances are made and the roles played by metaphor in shaping and directing the process of scientific discovery. In reviewing this little bit of scientific history we are also led to ask just what it means to say that something has been "discovered" in science.

As we saw in chapter 4, it was widely recognized in the late nineteenth century that matter is made up of atoms and that the atoms of different elements differ in various ways, such as their masses and chemical properties. Mendeleev published in Russia in 1869 a classification of the known elements, which he ordered in terms of increasing atomic weights. He noted that many properties of the elements appear to occur in a periodic fashion. Using his "periodic table," Mendeleev boldly predicted the properties of two elements as yet undiscovered and suggested where they might be found in nature. His periodic table served as a powerful guide in classifying new elements and had significant predictive power. But none of these chemically important advances in classification bore directly on the question of whether atoms, as tiny, discontinuous bits of matter, actually exist. In the late nineteenth century atoms remained nearly as elusive as they were to John Dalton, who had postulated at the beginning of the century that atoms are the smallest building blocks of matter that possess distinct chemical properties.

It had been observed that under appropriate conditions, matter has electrical properties. Alessandro Volta in 1800 built the first device for producing an electrical current. His so-called voltaic piles, forerunners of modern batteries, provided a source of electrical energy that could be used to cause chemical reactions to occur. Water could be decomposed into the elements hydrogen and oxygen. Humphrey Davy used electrical energy to isolate many new elements, including the very reactive metals sodium and potassium. Michael Faraday later made quantitative measurements by inserting conducting rods (electrodes) into solutions containing dissolved salts and measuring the extent of chemical reaction at the electrodes from the passage of electrical current. Faraday's work suggested that in such systems a unit of electrical charge is carried by some sort of particle, to which George Johnstone Stoney in 1874 gave the name *electron*. Stoney estimated the charge of the electron to be about 10^{-20} coulombs, a very tiny amount of charge.

Physics was at the time a far from settled world, despite the view of some that all the most important discoveries had been made. It had been known since the mid-1850s that when an electrical charge is placed across a pair of separated metal plates inside a partially evacuated glass tube, the remaining gas becomes electrically conducting. "Rays" of some kind emanated from the negatively charged plate, called the cathode, and moved in straight lines across the tube toward the positive electrode, called the anode. The rays could be made visible by placing phosphorescent materials in their path. Under the right conditions, even the glass walls of the tube could be made to glow with visible phosphorescence. Cathode rays became the subject of many studies, and controversies arose over the interpretations of experimental results.

Interest in cathode rays was high in both Germany and Britain, and the research efforts in the two countries became intensely competitive. In the earliest experiments it was shown that the rays were blocked by a solid body placed in the tube in the space between the cathode and the anode. The rays were shown to travel in straight lines in the absence of any disturbing influence, but a magnetic field caused the rays to bend. Today these properties of the cathode rays are demonstrated regularly in introductory physics and chemistry classes, as shown in figure 5.1.

What could these cathode rays be? Only two metaphorical constructs seemed plausible: Either they were waves of some kind, akin to light waves, or they were charged particles. The fact that they were deflected by a magnetic field spoke in favor of their being charged particles, but a majority of German physicists subscribed to the wave hypothesis anyhow, suggesting that perhaps this was a new kind of wave, with heretofore unobserved properties. On the British side, C. F. Varley first suggested in 1871 that the rays might be particulate in nature; William Crookes proposed that they were molecules that have acquired a negative charge from the cathode and are repelled by it. A few years later the well-known German scientist Heinrich Hertz observed that the rays were not deflected when an electrical potential was applied to a pair of plates between which the cathode rays passed. This result pointed away from the particle hypothesis; the applied electrical field should have deflected a charged particle.

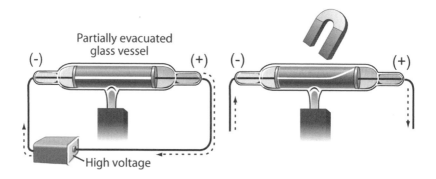

Figure 5.1. Bending of cathode rays by a magnetic field.

Joseph John Thomson began his undergraduate career in Trinity College at Cambridge University in 1876, at age 20. He started his research in the Cavendish Laboratory of Physics in 1880, under Lord Rayleigh. When Rayleigh resigned the Cavendish Chair in Experimental Physics in 1884, Thomson was elected to it, amid grumbling in some quarters that boys were being made professors. He began study of the electrical conductivity of gases and turned his attention to the properties of cathode rays in about 1890. In that year, Arthur Schuster, assuming that cathode rays were particulate in nature, calculated the ratio of charge to mass of the cathode rays based on their deflection in a magnetic field. Two years later, Hertz, still convinced that the rays are wavelike in nature, showed that they can penetrate thin foils of metal, in contrast to the results obtained earlier when thicker objects were placed in the beam's path. (The experimental results obtained in the study of cathode rays varied over time, largely because of improvements in experimental arrangements, particularly in the ability to produce higher vacuums in the tubes.)

So we have in Thomson a brilliant young physicist entering a research area widely thought to be important, where there are confused results, strongly conflicting alternative hypotheses, and something approaching a nationalistic competition between the British and German schools to arrive at a correct understanding.[2] Thomson opted for the particulate metaphor for cathode rays. That metaphorical model carried with it entailments that provided the basis for design of new experiments. If the rays are particles with charge and mass, they should be deflected by both electrical and magnetic fields. Thomson and his students showed that the earlier failure to detect deflection by an electrical field resulted from excessively high gas pressure in the tube. At appropriately low pressures, bending of the rays by both electrical and magnetic fields was observed. Thomson was able to design an apparatus in which both electrical and magnetic fields could be applied. The cathode rays could be formed into a narrow beam, which fell on a phosphorescent screen something like that on a TV tube or computer monitor. Figure 5.2 illustrates the apparatus. The electrical and magnetic fields caused the beam to deflect in opposite directions. Assuming that the rays are particulate, it was possible from the data obtained to calculate the speed of the electrons and e/m, the ratio of their charge e to mass m. This quantity proved to be a constant independent of the velocity of the rays. Thomson went on to show that when the cathode rays were captured in a metal vessel they gave up a negative electrical charge, thus establishing independently that they consist of charged particles. Thomson's experimental program illustrates how a model points the way toward new experiments that either confirm or disconfirm the model.

The value of e/m measured for the cathode rays was about 1,700 times larg-

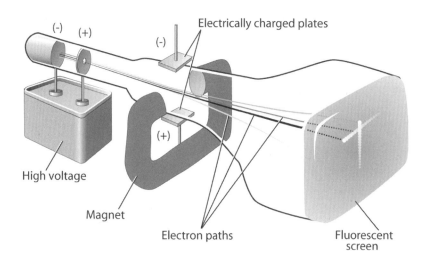

Figure 5.2. Thomson's electron beam apparatus.

er than that calculated from other experiments for the charged atom of hydrogen (the lightest of all atoms). It was possible to estimate the magnitude of the charge of the cathode ray particles at least crudely, and it proved to be about the same magnitude as that for the charged hydrogen atom. Without knowing anything definite about the nature of the charged hydrogen atom, Thomson could then conclude that the mass of the cathode ray particles is only about $\frac{1}{1,700}$ that of the hydrogen atom.

In succeeding experiments Thomson found that the properties established for cathode rays were quantitatively the same when different metals were used as the cathode and when different gases were present in the tube. Furthermore, he discovered that the same particles are emitted by metals when raised to a high temperature and by chemically very active metals, such as sodium and potassium, even when they are at room temperature. They were also found in abundance in the emissions of radioactive elements such as uranium and radium. The charged corpuscles he studied seemed to be present in everything.

A broader metaphorical construct was needed to incorporate these findings. It is embodied in the text of Thomson's 1897 paper in *Philosophical Magazine:* "We have in the cathode rays matter in a new state, a state in which the subdivision of matter is carried very much farther than in the ordinary gaseous state: a state in which all matter—that is, matter derived from different sources such as hydrogen, oxygen, etc.—is of one and the same kind; this matter being the substance from which the chemical elements are built up."[3] Thomson had leaped to the conclusion that the particles in the cathode ray tube, which others quickly identified with Stoney's electron, are present in all matter: "Thus, the atom is not the ultimate limit to the subdivision of matter; we may go further and get to the corpuscle, and at this stage the corpuscle is the same from whatever source it may be derived."[4]

A new and richer metaphor for electrons had emerged: The atom, believed for millennia to be the indivisible, smallest unit of matter, is in fact divisible. One of its components, present in all atoms, is a charged particle of low mass. It is in the context of this richer metaphor that Thomson can be said to have "discovered" the electron. He had moved beyond the realm of electrically conducting gases in evacuated tubes to a much more general search for charged particles with the properties he had identified for the cathode rays. The inquiry had broadened under the influence of a new metaphorical framework, one that grew in a natural way from the expanding scope of the study of cathode rays. Thomson reached a conclusion that seemed entirely logical: Electrons are a subatomic constituent of all matter.

J. J. Thomson's Atomic Model

Thomson's work raised a host of new questions about the nature of atoms. If the atom is composed in part of charged negative particles with mass, there must be some corresponding positive charge present to give the atom an overall neutral charge. But how many electrons are there in a typical atom? Is the positive charge also associated with particles of mass? If so, what are the relative masses of electrons and the positive charge? How are these charges distributed in the atom?

These and many other questions arose quite obviously from Thomson's results. Taken together, they begged for a model that might address them. The problem was, there were almost no experimental data that could shed light on the questions. One of the first models proposed, nearly simultaneously, by J. J. Thomson and Lord Kelvin, another British scientist, featured a uniformly distributed cloud of positive electricity into which the electrons were inserted to produce a balanced, equilibrium distribution of charges. In these models the positive charge was pictured as a continuous distribution of charge. Thomson initially hypothesized that the positive charge contributed very little to the mass of the atom. This meant that for the electrons to make up the mass of the atom there had to be large numbers of them—nearly 2,000 in the case of hydrogen, the lightest element. The Thomson model became known as the "plum pudding" model. The positive charge played the role of pudding, and the electrons were the plums embedded in the pudding.

Thomson's first model was a nonstarter, and it was not long before he revised it. A class of experiments called scattering experiments provided new insights into the internal structures of atoms. Directing a narrow beam of electrons, X rays, or light rays toward a sample produced scattering. The beam might just pass straight through, be absorbed in some way, or be deflected. The amount of deflection and the relative amounts of the beam deflected at various angles could be used to infer what the distribution of charge might be inside the atoms of the sample.[5] When Thomson analyzed the results of scattering experiments from various laboratories, he concluded that the number of electrons in the atom must be much smaller than he had originally proposed. This in turn led to certain obvious conflicts of the model with the laws of physics understood at that time. Thomson had proposed that the electrons were in motion in circular orbits within the atom. The laws of physics at that time indicated that such moving charges would radiate energy outward, and therefore the electrons could not be indefinitely stable in those orbits. However, because his revised model seemed to fit with the experimental results on scattering, Thomson simply ignored the discrepancies.

How was it possible for Thomson to adhere to a metaphor that seemed so much at variance with the laws of physics? The answer seems to be that Thomson felt intuitively that the study of subatomic structure was taking physics into new territory. He was prepared to find that some of the familiar laws of physics would ultimately prove not to be applicable. For example, although it seemed common sense that there would have to be positive charge in the atom to balance the negative charge of the electrons, he was not convinced at one point that the positive charge would ultimately be needed at all. Given this outlook, it made sense to stick with a model that seemed to map reasonably onto available experimental data, even though it might have potentially fatal flaws. This example highlights the role of metaphors as tools in discovery. They can continue to point the way toward new experiments even as their apparent failings call attention to the need for improvement or replacement.

By 1910, Thomson's plum pudding model of the atom had taken center stage, but many others had been advanced since the discovery of the electron in 1897. For example, in 1901, Jean Perrin, whose studies of Brownian motion were related in chapter 4, suggested a purely speculative model based on the analogy with the solar system:

> Each atom will be constituted, on the one hand, by one of several masses very strongly charged with positive electricity, in the manner of positive suns whose charge will be very superior to that of a corpuscle, and, on the other hand, by a multitude of corpuscles, in the manner of small negative planets, the ensemble of their masses gravitating under the action of electrical forces, and the total negative charge exactly equivalent to the total positive charge, in such a way that the atom is neutral.[6]

Perrin's model illustrates how experience with the macroscopic world is brought to bear in reasoning about the microscopic world beyond direct observation. That Perrin should have suggested this model very soon after the discovery of the electron, before there was any significant evidence bearing on the internal structure of atoms, is testimony to the active role of embodied reasoning based on direct experience. As it turned out, his model was lost sight of, only to be reinvented a decade or so later. In a similar vein, the Japanese physicist Hantaro Nagaoka drew on a well-known paper of James Clerk Maxwell that purported to account for Saturn's rings. He proposed in 1904 an atomic model in which the electrons revolve in rings around a positively charged body. Nagaoka's model had a fatal weakness: Repulsions between the negatively charged electrons should produce instability. Nevertheless, it illustrates once again the tendency of scientists to look in the observable physical world for models of microscopic systems.

The Nuclear Atom

The hero of the next stage of our story is Ernest Rutherford, a farmer's son born in 1871 in New Zealand. Ernest had eleven siblings. As a young boy of ten he was inspired by a popular physics text and determined that he would, if possible, make his way as a scientist. He attended Canterbury University College in Christchurch, New Zealand. On the basis of his work there, which included two research papers published in the *Transactions of the New Zealand Institute,* he won a scholarship to Cambridge. Rutherford was admitted in 1895 to a new Ph.D. program at Cambridge, designed for students who had done their undergraduate work elsewhere. He met J. J. Thomson and became one of his first Ph.D. students.

Rutherford's scientific career was spectacular. Within two years he had completed his doctoral research. A year later, supported by Thomson, he was offered a professorship in physics from McGill University in Montreal, which he accepted. During his time at McGill he studied radioactive emissions, identifying the alpha, beta, and gamma emissions as three distinct forms. The alpha rays were found to be positively charged and to have mass. Thus, they were particulate in nature, as were the beta rays, which were already established to be electrons. The gamma rays were a form of high-energy radiation akin to X rays. The work on radioactivity carried out at McGill earned him the Nobel prize in 1908 at the young age of 36.

Rutherford returned to England in 1906 to take up the physics chair at Manchester University. While at McGill, he had shown that the positively charged alpha particles, emitted from radioactive elements such as radium, are much more massive than the beta ray particles, which were known to be electrons. At Manchester he began a study of the scattering of alpha particles by matter. An early finding in the new laboratories was that the alpha particles were positively ionized helium atoms.

The studies of scattering of alpha particles by matter culminated in what was arguably the most important work of Rutherford's career. A new student in the laboratory, Ernest Marsden, was set to work under the direction of Hans Geiger, a research assistant, to examine the scattering of alpha particles by air. The results were not easily interpreted. Rutherford made a suggestion that seemed at the time to be a bit off the wall: Put a thin metal foil in front of the beam of alpha particles and look for scattering of the particles at large angles. In other words, look for some of the alpha particles to be deflected off to the side or even to be reflected back in the direction from which they had come. Assuming the Thomson plum pudding model of the atom, it was not expected that the alpha particles, known to be several thousand times more massive than the electron and moving at high speed, would be scattered significantly in passing into and through the foil. A rough analogy might be the passage of a rifle bullet through a large cardboard box filled with cotton balls and lead buckshot widely distributed throughout the cotton. The bullet might encounter several of the low-mass buckshot on its way through the box, but even multiple encounters would not deflect it much. On the other hand, Rutherford also knew that alpha particles showed a much lower capacity to penetrate matter than did electrons. It wasn't clear why this should be so if the plum pudding model was correct.

The results of Marsden's experiments led to one of the great milestones in our understanding of the nature of matter. He found that a small fraction of alpha particles were deflected at large angles; about one in 20,000 were deflected 90° or more. In popular accounts Marsden's report to Rutherford is portrayed as one of those "eureka!" episodes in which the concept of the atom underwent a sudden, wrenching change. There is no real evidence that such an epiphany occurred. The results were certainly anomalous and were reported as such by Marsden and Geiger and then by Rutherford in subsequent presentations during 1909 and 1910. However, it is significant that the scattering model based on Thomson's plum pudding model was not immediately abandoned. More than a year passed before Rutherford announced to Geiger during the winter of 1910–11 that he knew what the atom looked like and how to explain the large deflections of the alpha particles. The famous paper describing his concept of the nuclear atom appeared in the spring of 1911.[7] In it, Rutherford describes a theory of alpha particle scattering based on the idea that the atoms of the metal foil consist of a tiny but massive nucleus that contains the positive charge of the atom. The electrons surround the nucleus and take up nearly all the volume of the atom.

From the historical record we can gain insights into the role of metaphor in the evolution of Rutherford's reasoning about the scattering results. As already mentioned, the Geiger-Marsden data did not immediately cause Rutherford to abandon Thomson's plum pudding model. Nevertheless, the scattering behavior of the alpha particles was anomalous. The reasoning goes something like this: Let's imagine the alpha particle, which is known to have the mass of a helium atom, moving into a gold foil, and imagine that the gold atoms are constructed in accordance with the plum pudding model. This means that the electrons, of much lower mass than the alpha particles, are in circular orbits of some kind, and the positive charge is uniformly distributed throughout the atom. The alpha particle in this model will undergo numerous encounters with charged particles, but because no one of them, positive or negative, is presumed to have a mass nearly as large as the alpha particle, each individual encounter can cause only a small deviation in its path. However, there will be many such encounters; the net result is that the observed deflection will be the result of many individual small deflections, in one direction, then in another, and so on. From this physical picture, one can formulate a mathematical description. That model predicts that the chance of a deflection at angles as large as 90° is negligibly small.

Rutherford's notes and discussions with colleagues reveal that over a period of a year or so, he came to the conclusion that multiple scattering of the alpha particles could not account for the Geiger-Marsden data. He turned then to the idea of a very small, massive, charged particle at the center of the atom. The volume of the atom is mostly taken up by the electrons, which have a low mass. When the positively charged alpha particles, with a mass several thousand times that of the electron, pass into the metal foil at high speed, they mostly encounter electrons. But the electrons are so low in mass compared with the alpha particles that they do not produce deflections. Thus, most of the alpha particles pass through the thin metal foil, which is several thousand atomic layers thick. There is a small probability that the alpha particle will have a head-on collision with or pass very close to one of the tiny nuclei. Should this happen, there will be a strong electrical repulsion between the two particles because they are both positively charged. Gold atoms are about fifty times more massive than alpha particles. Assuming that nearly all this mass is in the tiny nucleus, the alpha particles do most of the movement in avoiding a collision. They veer off, sometimes at large angles. Because the nucleus occupies a very small volume, only one in about 20,000 of the alpha particles has such a near-collision. Rutherford was able from this physical model to derive a formula for the angle dependence of the scattering that agreed reasonably well with experiment.

It is evident that the formulation and interpretation of these alternative metaphors incorporated a great deal of intuitive physical understanding on Rutherford's part. The models, highly geometric in character (Rutherford's notes are replete with sketches), formed the basis of mathematical descriptions that in turn could be tested against experimental data. It is important to note once again that the mathematical descriptions are themselves metaphors. This is so because they are no more than descriptions in mathematical language of the model constructed to account for the observations. They are another form of what might be said verbally or diagrammed.

Rutherford's paper created hardly a ripple. J. J. Thomson continued to lecture on his plum pudding model. When he did mention Rutherford's work in a presentation two years after the paper had appeared, he disagreed with Rutherford's interpretation. He preferred to believe that there might be some special kind of interaction between the incident alpha particles and alpha particles that Thomson thought could be present within the atoms. Like everyone else, scientists often adhere to cherished beliefs in the face of strong conflicting evidences. Rutherford, who was given more to reticence than enthusiastic self-promotion, contented himself with continuing experimental studies that might shed further light on the model. But it happened that a young Danish scientist, Niels Bohr, fresh from his doctoral work at the University of Copenhagen, arrived in England in 1911 with a one-year fellowship. Bohr went first to Cambridge to work with the great J. J. Thomson. Things did not go well for some reason. In the wake of rather frustrating experiences at Cambridge, he was happy to have the opportunity to spend four months with Rutherford in Manchester, beginning in the spring of 1912. Once settled in, Bohr turned his attention to Rutherford's nuclear model of the atom. Before long he was ready to announce a radically new metaphor for the atom, based on quantum theory.

The Quantum Atom

Rutherford's 1911 paper on the nuclear atom, successful as it was in accounting for the alpha particle scattering results, was afflicted with a major shortcoming: It did not explain how an assembly of electrons circulating around a central charge could be stable. According to classical theory, a charged particle in a circular orbit should radiate energy. The energy it radiates must come from the energy of its motion around the nucleus. As it loses energy through radiation, the electron should simply spiral downward, in smaller and smaller orbits, until it collapses into the nucleus. It is perhaps because this problem was not addressed that Rutherford's model received only lukewarm reception. It was just this problem that Bohr had to solve.

The theoretical physics community was at the time abuzz with talk of the significance of something called quantum theory. In 1900 a German physicist, Max Planck, had proposed a radically new theory to explain how bodies radiate energy. Any body, such as a red-hot coal or a hot biscuit fresh from the oven, emits energy in the form of electromagnetic radiation. Some of this radiation we can feel as heat (infrared) rays or see as visible light. There was at the time no satisfactory theory to account for the various kinds and amounts of radiation emitted by a body at any particular temperature. Planck proposed the introduction of a new entity, called the quantum of energy. The essence of Planck's idea, as it eventually came to be understood, is this: When an object is emitting energy (e.g., as the hot biscuit emits infrared heat energy), the energy is not given off in entirely arbitrary amounts, large or small, at one time. Rather, it is given off in discrete chunks, called quanta. A rough analogy might be as follows: Think of a dump truck filled with sand. The sand can be dribbled off the back of the truck in any amount, large or small, even as little as one grain. This picture corresponds roughly to the classical way of emitting energy, in arbitrarily large or small amounts. Now suppose that the truck is loaded with sand that has been compressed into bricks. It can lose sand in units of bricks, but nothing smaller than one brick. This corresponds to the quantum model of emitting energy. In this analogy the "quantum" of sand is one brick.[8]

Planck's quantum theory of radiation was successful in accounting for the distribution of radiant energy in the normal spectrum of radiant heat from a body. His theory, expressed in mathematical form, contained a constant, h, which Planck called the elementary quantum of action. The smallest amount of energy that could be emitted by an object was determined by the product $h\nu$, where ν represents the frequency of the radiation. But what did the theory really mean? By what process of reasoning did Planck light upon the concept of the quantum? How and in what ways were Planck's thought processes driven by reasoning based on everyday physical experience? Was his solution more than merely a fix to a mathematical formulation?

It is true that Planck came upon the mathematical formulation by simply noting that it would be necessary to have a new term in the radiation formula, with a new constant, to fit the data. But having found this solution, he was not thereby satisfied: "But even if this radiation formula should prove to be absolutely accurate it would after all be only an interpolation formula found by happy guesswork, and thus would leave one rather unsatisfied. I was, therefore, from the day of its origination, occupied with the task of giving it a real physical meaning. . . . After some weeks of the most intense work of my life clearness began to dawn upon me, and an unexpected view revealed itself in the distance."[9] Planck's experience exemplifies yet another pathway to scientific discovery. Mathematical formulations of theories and laws are forms of description. The scientist can use these descriptions as reasoning devices to achieve, sometimes on entirely empirical grounds, more satisfactory fits of the description with the observables at hand. This process may occur without using metaphorical reasoning to any great extent. However, the absence of a more physically based model may leave the scientist with the feeling that something is missing, as was apparently the case with Planck. The view that opened itself to him had to do with certain principles of statistical physics, with which we need not concern ourselves. It was not a truly satisfying view, and Planck remained uneasy with the radical assumption he had made in solving the radiation problem. It lent itself to no clear metaphorical image of a mechanism operating in the physical system.

Thomas Kuhn argues convincingly that Planck did not recognize that his formulation demanded a restriction on the classical model of radiation.[10] It remained for Albert Einstein in 1905 to invoke the quantum concept in a way that generated a powerful physical image. It was known that when radiant energy—that is, light rays—fall on certain metal surfaces, the metal surface may emit electrons. The effect, called photoemission, is the basis of the "electric eye" cell, used in door openers and light dimmers. The diagram of a phototube, shown in figure 5.3, illustrates the experiment. Light enters the tube and falls on the

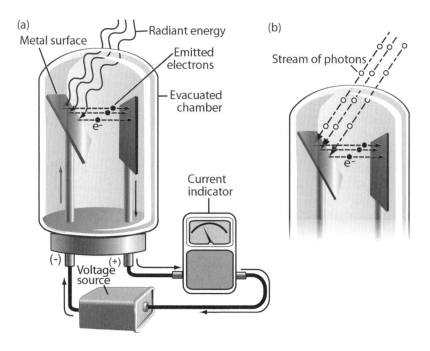

Figure 5.3. Photoemission of electrons from a metal: (a) the classical view of radiant energy as a wave phenomenon; and (b) in Einstein's model the incoming radiation is viewed as a stream of photons.

metal surface. If electrons are emitted, they are drawn to the positively charged plate in the tube, and a current is seen to flow. Experimental studies using such a cell had produced results that could not be satisfactorily explained by the theory current at the beginning of the twentieth century.

Einstein made a bold move: Taking his cue from Planck's quantum theory of radiation, he replaced the classical view of the incoming radiation as a continuous wave, shown in figure 5.3(a), with one in which the radiation is composed of little packets of energy, which he called photons, as illustrated in figure 5.3(b). Each photon has an energy $h\nu$. In other words, in Einstein's model, the radiation has been quantized. The surface is not absorbing a continuous wave but rather is being bombarded by a stream of massless energy bullets. Each photon transfers its energy to an electron. If the energy of the photon is high enough, the energy imparted to the electron enables it to escape the attractive forces of the metal, and current flows. If it is not sufficiently energetic, the energy is merely dissipated in the metal as heat, and no emission occurs. Photons of sufficiently high energy—that is, with sufficiently high-frequency ν—cause emission. Those of lower energy do not. Using the model of quantized radiation, Einstein was able to account satisfactorily for the extensive observations relating to the photoelectric effect. The 1905 paper was a tour de force. It earned him the Nobel prize in 1921.[11]

Niels Bohr recognized that the problem of the atom was, in some ways, analogous to the problems of emission of radiant energy and of photoemission. The classical picture doesn't work because it is based on the idea of a continuous range of values for energy on the atomic scale. Planck and Einstein had shown that one must think in terms of discrete, or quantized, amounts of energy. Bohr formulated a model for the simplest atom of all, hydrogen, which has but one electron surrounding the nucleus. If the electron is in a stable, or stationary, state, the classical rule about radiating energy does not apply. These stable states are ones in which the angular momentum of the electron (its mass times its angular velocity) is quantized. That is, the angular momentum can have only values that are integral multiples of $h/2\pi$ (i.e., $h/2\pi$, $2h/2\pi$, $3h/2\pi$, etc.). This means that the electron can orbit the nucleus only at certain specific distances from the nucleus.

Bohr's model matched many observed properties of atomic hydrogen. Most impressively, it accounted for hydrogen's spectral properties. When atoms are placed in an electric discharge, they absorb energy from the electrons in the discharge and eventually emit light when they release this energy in returning to their original state. We see this most commonly in neon lights and in sodium and mercury vapor streetlights. The radiation emitted from an element such as hydrogen consists of only certain specific wavelengths. The classical theory had no way of accounting for these observations. In Bohr's model they represented transitions of the electron from one allowed, or stationary, state to another. Because the energy of each state is sharply defined by the quantum condition, only specific differences in energies of the allowed states exist, and therefore only specific values can be seen for the wavelengths of the emissions.

The advent of quantum theory seems to spell trouble for the idea that science proceeds using models based on embodied experience with the macroscopic world. Planck's imposition of a quantum condition on radiation was an empirical move; he used it because it worked. There is no evidence that Planck had any notion of the sort of physical character of photons that Einstein later gave them. By contrast, Einstein's treatment of radiant energy as a stream of quantized energy particles is readily visualized in terms of a physical metaphor. Bohr was drawn to his model of the quantum hydrogen atom by thinking in terms of lengths and magnitudes.[12] He saw that a successful model for the nuclear atom must be one that allows the electrons to exist in orbits of atomic size. That is, the electrons must move about in such a way that they define an atom on the order of the sizes estimated for atoms from various experimental studies. Once the quantum condition has been imposed, the model of the atom in an allowed state is no different from the classical planetary picture of the electron orbiting the nucleus. However, a difficulty arose in the lack of suitable metaphors for the transition of the atom from one allowed state to another. When Rutherford read the papers Bohr submitted to him in 1913 for publication in *Philosophical Magazine,* he had great difficulty in assimilating the idea of a transition from one allowed state to another, the key element in Bohr's explanation of the spectral properties. He wrote to Bohr, "There appears to be one grave difficulty in your hypothesis, which I have no doubt you fully realize, namely how does an electron decide what frequency it is going to vibrate at when it passes from one stationary state to another?"[13] We see in this query that Rutherford is thinking in terms of the location metaphor for a change in state and imagining the electron to be in transit between the states for a time sufficiently long to permit one to talk about its vibrations. But in quantum theory the transition occurs essentially instantaneously; there is no clear physical model for the transition process itself.

Quantum theory created a sense of crisis in physics because it meant that the behavior of atomic- and subatomic-level systems could not be understood entirely in terms of metaphors drawn from experience with the macroscopic world. Bohr put it this way: "We know from the stability of matter that Newtonian physics does not apply to the interior of the atom; at best it can occasionally offer us a guideline. It follows that there can be no descriptive account of the structure of the atom; all such accounts must necessarily be based in classical concepts which, as we saw, no longer apply."[14]

Particles and Waves

There is no need to trace all the transformations through which the concept of the atom progressed to bring us to the present. However, to appreciate the nature of most recent experimental work dealing with atoms we must understand yet another revolution in the conception of matter at the microscopic level. We have seen that Einstein introduced the notion of radiant energy as a stream of massless energy particles, the photons, moving at the speed of light. Because each photon has energy defined by the frequency of the light, each photon can be thought to have momentum, analogous to a moving object such as a bullet or a rolling billiard ball. When such an object strikes some other object, it transfers momentum. For example, when a billiard ball strikes another billiard ball, its own speed and direction of motion are changed. In a similar way, when a photon strikes a sufficiently small object, such as an electron, it may undergo a change in momentum. This simple physical analogy provided the basis of an experiment by Arthur Compton: He showed that X ray photons could be scattered by electrons with a change in the momentum of the photon, just as predicted on the basis of this metaphor drawn from experience with the macroscopic world. But the homely, familiar character of the metaphor cloaks a great mystery: How can radiant energy behave like a particle?

The metaphorical view of radiant energy as wavelike is supported by many experimental observations, such as the refraction of visible light into rainbows. Yet, as Einstein showed, this same radiant energy behaves like a stream of particles in producing photoemission. If radiant energy, with its clearly wavelike character, can under some circumstances behave like a particle, might not particles of sufficiently small mass under certain circumstances behave like waves? So reasoned a young French physicist, Louis de Broglie, who bet on the idea in a big way when he defended it before his doctoral thesis committee in 1924. His argument was bolstered by complex mathematical development, including relativity theory, but ultimately it rested on the simple analogy just described. The committee was reported to be reluctant and skeptical, but they passed him. It was as well they did so; his doctoral thesis was the basis of his Nobel prize only five years later.

De Broglie's idea of wave-particle duality can be applied to the electron in the hydrogen atom. Think of the electron orbiting the nucleus as a wave rather than a particle. It turns out that the circumference of the electron's path in one of Bohr's allowed orbitals is exactly a whole number multiple of the wavelength as calculated by de Broglie. This means that the wave pattern is stable, or "allowed." Thus, there was consistency between de Broglie's theory and Bohr's much-discussed application of quantum theory. Not only that, only a few years after de Broglie's thesis defense, C. W. Davisson and L. H. Germer in the United States and G. P. Thomson in England demonstrated that an electron beam directed into a crystal is scattered in just the same way as X rays.[15] Under the right circumstances the electron, which J. J. Thomson went to great lengths to show is a particle, behaves as a wave.

If the electron in an atom has a wavelike character, what is the atom really like? And what kind of model can we have of atoms other than hydrogen, which have more than one electron surrounding the nucleus? To add to the difficulties of answering this question, Werner Heisenberg announced in 1925 a theory of the hydrogen atom that brought forth a new, troubling consideration. He pointed out that there is a limit to the degree of precision with which we can simultaneously know the position, speed, and direction of motion of any object. This limit has no consequences for ordinary macroscopic objects, but for a tiny object such as the electron, it means that there is no way in principle to ever know the precise location of an electron in an atom if we know its speed and direction of motion. Heisenberg's uncertainty principle has been much discussed by scientists and philosophers. Among other things, it has been used to rescue free will from the clutches of classical determinism. We will not venture into those philosophical quicksands. Of more interest for our purposes is Heisenberg's assertion that the uncertainty principle renders conventional models of the microscopic world of the atom useless. Only a mathematical model can be taken as the basis of useful reasoning about such matters as atomic structure.

Heisenberg offered a mathematical method for calculating the properties of atoms containing more than one electron. His so-called matrix mechanics was deemed impenetrably difficult. Einstein wrote, "The calculation is pure witchcraft. . . . It is most ingenious, and owing to its great complexity, safely protected against any attempt to prove it wrong."[16] Erwin Schrödinger put forth a more comprehensible model in 1926. It combined the elements of de Broglie and Bohr in a new formulation of the electronic structure of the atom, which he called wave mechanics. We need not go into the details of his heavily mathematical development. The essence of Schrödinger's approach was to treat the electrons in the atom as waves that obey certain rules, so-called boundary conditions. What comes out of the theory is a series of mathematical functions called orbitals. Each orbital corresponds to a possible state of the electron in the hydrogen atom. The various orbitals possible for hydrogen match exactly the allowed energy states for hydrogen calculated by Bohr. The difference in the two models is that, whereas in the Bohr atom the electron is viewed as a particle moving in a circular orbit, in the Schrödinger model the electron is a wave distributed in space in a certain way. Max Born pointed out in 1926 that the square of Schrödinger's mathematical function is a measure of the relative probability that the electron will be found at a particular point at any given instant. Over time, the electron is most likely to be located where the probability is large.

Born's idea was not well received by Schrödinger, among others. Such an interpretation gives rise to deep questions regarding the meaning of measurements of the properties of subatomic particles. The debates, which continue to the present, engaged the passionate participation of most of the big names in theoretical physics, including Einstein and Bohr. These two sustained an extended dialogue on the ultimate viability of quantum theory as a satisfactory model of nature.

Aside from the debates over philosophical issues, Schrödinger's wave mechanics has been highly successful as a model of the electronic structure of atoms. In one form or another it remains the fundamental model on which atomic structure is modeled. Although the theoretical formulation of the model can make for difficult going, the results have a clear physical meaning.[17] A theorem called the Hellmann-Feynman theorem states that the electron distribution calculated from Schrödinger's wave function corresponds exactly to a classical distribution of electrical charge. That is, the electrons in the atom can be thought of as a static cloud of negative electrical charge, with the total charge equal to the charge of all the electrons in the atom. The places where the electrical charge cloud is thickest are those where Schrödinger's function has its largest values.

This correspondence between the charge distribution calculated from wave mechanics and a classical distribution of charge enables us to represent the atom as a nucleus surrounded by a cloud of negative electrical charge, as shown in figure 5.4(a). The picture can be extended to molecules, which are made up of atoms chemically bonded together, as illustrated in figure 5.4(b).

In the thirty-year period extending from J. J. Thomson's discovery of the electron in 1897 through Erwin Schrödinger's presentation of his wave mechanical model of electronic structure within the atom in 1926, scientists' concepts of

(a)   (b)

Figure 5.4. Image of the distribution of electronic charge: (a) in the hydrogen atom; and (b) in the hydrogen molecule, H_2. The nuclei are shown as small, centrally located spheres; the electrons are shown as a "cloud" of distributed electrical charge.

the nature of the world on the atomic scale underwent truly revolutionary change. We have focused only on selected episodes of importance for metaphorical understandings of the nature of atoms. Einstein's development of special and general relativity theory also occurred during this time. It was also discovered that, in addition to positively charged mass units called protons, the nucleus contains neutral particles of the same mass, called neutrons. Not all atoms of an element need have the same mass because the number of neutrons in the nucleus may differ. (Atoms of a given element that differ in mass are called isotopes.) Rutherford and his colleagues discovered the transmutation of elements through radioactive decay. In the crowning achievement of a remarkable scientific career, Rutherford later showed that it was possible to artificially induce transmutation of one element into another. These discoveries, important as they were, did not have major impacts on the metaphorical conception of atoms that most concerns us.

At the beginning of the 1897–1926 period, the prevailing metaphor for the atom could fairly be called the billiard ball model: that of an indivisible, immutable entity. By 1926 the model that had gained ascendancy was based on a complex mathematical development that only a few nonscientists, or scientists for that matter, could follow. This development poses a challenge to our understanding of the role of metaphor in scientific discovery. In what sense can reasoning based on gestalts formed from direct experiences with the physical world be said to have played a role in the invention of quantum theory, the photoelectric effect, Bohr's model of the hydrogen atom, de Broglie's theory of wave-particle duality, Heisenberg's uncertainty principle, or Schrödinger's formulation of wave mechanics?

It could be argued that the break with classical physics needed to formulate satisfactory theories meant that metaphors based on experiences drawn from the macroscopic world, where quantum effects are not discernible, could no longer apply. In his book *Physics and Beyond,* Werner Heisenberg describes many intense and prolonged discussions between himself and Bohr, Schrödinger, Born, Einstein, and others.[18] These great men, who had created an entirely new physics, were troubled by the realization that concepts that can describe events in daily life were proving inadequate for expressing the content of the new theoretical models, grounded in mathematics.

However, although Heisenberg's famous uncertainty principle came about as a purely algebraic consequence of his matrix mechanics theory, it is also the case that when he published his theory, Heisenberg introduced the uncertainty principle via a gedankenexperiment, the so-called microscope experiment. Heisenberg's thought experiment is a metaphorical construct based on analogy with a macroscopic physical model of one particle striking another and transferring momentum. Whether such a metaphor played any role in Heisenberg's recognition of the uncertainty principle we do not know. But it is clear that to put the uncertainty principle into terms that have meaning for most people, a simple physical metaphor is needed.

In a similar way, Schrödinger's wave mechanical model might be thought to have little relation to a physical metaphor. However, key elements of the mathematical formulation derive from a physical model of the system. For example, one of the requirements of the theory is that the wave function must tend to zero value at large distance from the nucleus. The aptness of this "boundary condition" depends on a model of the system in which the electron is attracted to a central nucleus. Furthermore, certain results of theory—for example, the shapes and extents of the computed electron clouds—find useful expression as three-dimensional models such as those shown in figure 5.4. Thomas Kuhn has this to say about the evolution of atomic theory during the twentieth century:

> Bohr and his contemporaries supplied a model in which electrons and nuclei were represented by tiny bits of charged matter interacting under the laws of mechanics and electromagnetic theory. That model replaced the solar system metaphor but not, by doing so, a metaphor-like process. Bohr's atom model was intended to be taken only more-or-less literally; electrons and nuclei were not thought to be exactly like small billiard or Ping-Pong balls; only some of the laws of mechanics and electromagnetic theory were thought to apply to them; finding out which ones did apply and where the similarities to billiard balls lay was a central task in the development of quantum theory. Furthermore, even when that process of exploring potential similarities had gone as far as it could (it has never been completed), the model remained essential to the theory. Without its aid one cannot even today write down the Schrödinger equation for a complex atom or molecule, for it is to the model, not directly to nature, that the various terms in that equation refer.[19]

The formulation of mathematical theories directed toward achieving a fit with experimental observations, without obvious reliance on a physical model, as in Planck's formulation of the radiation law, certainly was important during this revolutionary period. Nevertheless, it is also clear, often from the assertions of the scientists themselves, that physical models based on experience of the macroscopic world played a guiding role.

Seeing Atoms

Schrödinger's wave mechanical model remains the reigning theoretical metaphor for the atom. Put to work as a computational tool, it accounts for a host of properties of atoms of all the elements. Beyond this, it enables one to account for the structure of the periodic table and the chemical properties of the elements. But the picture it leaves us of the atom, or of molecules formed from atoms, is still the unsatisfactory one of probability distributions of electronic charge in the space around the nuclei. To further compound the ambiguity, Mylneck and coworkers in 1991 showed that helium atoms could be diffracted by a crystal as though they were waves! The helium atom is not a single particle; it is composed of a nucleus containing two protons and two neutrons, surrounded by two electrons. Yet under the correct experimental conditions this compound entity is diffracted by a lattice in the same way a wave would be. In principle, this experiment could be performed with a beam of atoms of any element, although it is in fact very difficult to perform. One lesson we can take from this is that *there is no single correct, objective, literal image of an atom.* Each image we may hold is the result of our metaphorical interpretations of data gathered in particular kinds of experiment. One set of experiments can be best understood in terms of a model in which the atom is conceptualized as a wave phenomenon. Other observational data can be interpreted satisfactorily only by thinking of the atom as an entity with a massive nucleus and orbiting electrons.

On the experimental side, there have been great advances in our ability to image individual atoms or molecules. But how real is our vision of the atom? Let's examine some of these visualization tools, with the aim of asking what it means to "see" matter at the atomic level. One of the earliest tools for studying the structure of matter at the atomic scale involved the use of X rays. W. C. Roentgen discovered in 1895 that very-short-wavelength (high-frequency, high-energy) radiation was emitted when electrons at high voltage were allowed to strike a metal electrode. These X rays produced a sensation when Roentgen made a photographic plate showing the bones of his wife's hand. Some years later it was realized that, because of their very short wavelength, X rays could be diffracted by a lattice of atoms in a crystal, just as visible light is diffracted by a diffraction grating. When X rays impinge on a single crystal of a solid substance, scattering of the rays occurs such that a photographic plate a short distance in back of the solid exhibits a highly symmetric pattern of spots distributed at various angles from the incident beam. In 1912–13 W. H. Bragg and his son W. L. Bragg showed that the complex pattern of spots could be interpreted in terms of a unique structure that must have given rise to the pattern. They reasoned backward from the pattern of the diffraction to deduce the kind of atomic lattice that must have caused the pattern. Their early studies were of simple solid structures such as sodium chloride, shown in figure 5.5. The figure illustrates only a small portion of the sodium chloride lattice, which goes on repetitively in all three dimensions.

The Braggs' work was tremendously important. It lent a sense of reality to

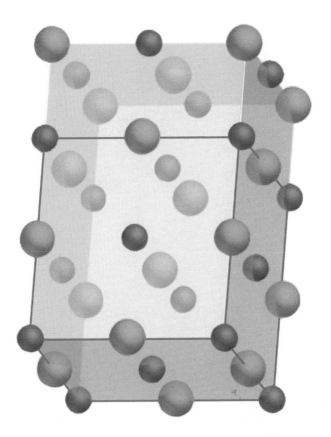

Figure 5.5. A portion of the lattice of sodium and chloride ions in
sodium chloride. (Reprinted with permission from Pearson Education, Inc., Upper Saddle River, N.J., from T. L. Brown, H. E. LeMay,
and B. E. Bursten, *Chemistry: The Central Science*, 9th Ed. [Upper
Saddle River, N.J.: Prentice Hall, 2003], 433, fig. 11.35B. © 2003.)

the idea of atoms and ions and provided a tool for modeling the three-dimensional arrangements of atoms and ions in crystals. X ray diffraction and related
diffraction tools have since become one of the most powerful tools in structural chemistry and biology. When a new chemical substance is produced in the
research laboratory, determination of the three-dimensional structure using X
ray diffraction is a routine procedure. In biochemistry, X ray diffraction data led
to beautiful pictures of large, complex molecules such as proteins. More than a
few Nobel prizes have rested on the determination of structures using X ray
diffraction.

Yet another powerful method for "seeing" atoms, the scanning tunneling microscope (STM), was invented by Gerd Binnig and Heinrich Rohrer during the 1980s. Their work caused an immediate sensation, and they were awarded the Nobel prize in 1986. It is at first blush something of a miracle that the STM experiment works at all, considering the exquisitely small dimensions involved and the level of control needed to carry out the experiment successfully. A simple diagram of the heart of the apparatus is shown in figure 5.6. The idea is this: An extremely fine tip of tungsten metal is brought to within a few atomic distances of a conducting surface, such as a metal, and an electrical potential is applied. According to classical theory there should be no current flowing between the tip and the surface. However, quantum theory predicts that there can be a small current, which results from the ability of the electrons to "tunnel" across the gap between the tip and the surface. The magnitude of the tunneling current depends very critically on the distance d shown in figure 5.6.

Now the tip is moved across the surface, and the tip is raised and lowered so as to maintain a constant current. Where the surface is higher or lower, the tip must be raised or lowered by tiny amounts to keep d constant. The variation in vertical position as the tip is moved across the surface is recorded. Then the tip is brought back to the starting point, moved to the side an atomic dimension or so, and the scan repeated. In this way a kind of topographic map of the section of the surface is created, based on the vertical positioning of the tip. The current data are processed using computer programs to produce pictures of surfaces that show atomic-level details. In recent years it has even become possi-

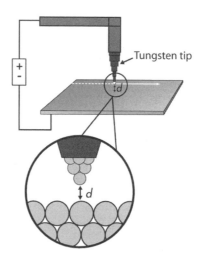

Figure 5.6. A schematic view of the STM.

ble to use the tip to pick up and move atoms one at a time, to arrange them on a surface in a desired pattern, as illustrated in figure 5.7.

What can we say about the status of the images of atoms that are produced from the X ray diffraction or STM experiment? Do they represent images of atoms in the same sense as our direct vision of, say, a tennis ball, or even a photo of a tennis ball? We have seen in earlier discussion that humans can extend the range of basic sensory perception with the aid of tools of various kinds. For example, by use of the telescope, Galileo discovered the moons of Jupiter, which are not visible to the unaided eye. Recall, however, that some were reluctant to credit Galileo's evidence. The telescope faithfully enlarges images of terrestrial objects, but can it be counted on to provide a faithful view of things in the heavens? The skeptics were not convinced that the same laws of nature that operate on Earth operate in space. So there can be a kind of skepticism that operates with respect to any instrumental extension of our senses: Does it provide a faithful representation of the physical system it purports to image?

This issue has certainly been put to rest for the optical telescope. Telescopic images may have to be corrected for certain imaging errors, just as our eyesight may need correcting by glasses or contact lenses. Nevertheless, from long experience with optical telescopes, we can say that the objects observed with them have the same ontological status as those we view with our unaided vision. This

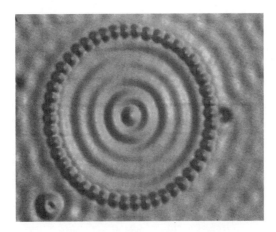

Figure 5.7. An STM image of 48 iron atoms arranged on a copper surface. The diameter of the ring is about $\frac{1}{20,000}$ the diameter of a human hair. (Reprinted with permission from M. F. Crommie, C. P. Lutz, and D. M. Eigler, "Confinement of Electrons to Quantum Corrals on a Metal Surface," *Science* 262 [1993]: 219, fig. 2. © 1993 American Association for the Advancement of Science)

claim can be extended to modern instruments that extend human vision not only in terms of magnification and sensitivity to low levels of light but also in terms of the portion of the spectrum of radiant energy that can be observed. The human eye is a weak instrument compared with the detectors on the Hubble space telescope.

When we consider our capacity to view microscopic objects, the same argument applies, up to a point. The typical optical microscope represents an extension of our vision in the same way as an optical telescope. But at some point in the push to see ever-smaller objects, simple optical methods no longer work. The wavelengths of visible light, and the properties of the materials from which microscope lenses are made, place limitations on the magnifications attainable. To go beyond this requires that we resort to an instrument that can produce very short wavelengths of radiant energy. One such instrument is the electron microscope, which takes advantage of de Broglie's wave-particle duality. We have already alluded to the discovery that an electron moving at a certain speed has associated with it a characteristic wavelength. A microscope can be built around this idea. By generating a beam of electrons moving at an appropriate speed, the electrons appear like very-short-wavelength radiation. The magnifications achieved with the electron microscope are very high, enabling scientists to image exceedingly small features of materials. But these images are seen only after processing of data from a complex instrument. Now the "lenses" of the microscope are electrodes with carefully controlled voltages and complex geometries. The instrument itself is based on the concept that electrons under appropriate circumstances may behave as waves.

With the X ray experiment the matter is more complex still. Here the observables are the angles at which the radiation is scattered and the intensities of the scattered X ray radiation, measured nowadays on very sensitive electronic detectors. To derive from these data information about the arrangement of atoms in a lattice, one must first use the theory of diffraction. But what are the scattering centers? We have the metaphor from wave mechanics of the atom as a cloud of electron density, surrounding the nucleus. X rays, it turns out, are scattered by the electrons and not the nucleus. But the precise way in which the scattering power of an atom depends on the number of electrons, the effects of chemical bond formation, vibrational motions of the atoms in the lattice, and other factors must all be taken into account. This taking into account involves metaphors, of course: models of scattering, thermal motion in the lattice, absorption effects, and so on. Finally, the relationship of the scattered radiation to the observed data involves a mathematical theory, also a metaphor.

What this means is that the beautiful images of atoms in lattices that we see as the output of X ray diffraction studies depend on multiple models. In the same way, the pictures we see from STM experiments, such as that in figure 5.7, depend on models, each of which is metaphorical in nature: a model for electron tunneling, with its quantitative expression in the form of a theory of tunneling current; a model of how the current relates to the nature of the surface being scanned; and complex computer programs that interpret the data and convert them to pictures that please the eye. For example, the color in STM pictures is added for visual effect; it is not inherent in the data.

To return to the original question of whether we "see" atoms, there are two levels of response. First, we must remember that any model we might use to characterize the atom is metaphorical, whether it be that of a billiard ball, a plum pudding, a miniature solar system, a cloud of negative charge surrounding a positive center, or a densely mathematical description based on quantum theory. Our experimental attempts to see the atom as it is all involve approaches that relate observables to the atom via one or more models. Thus, the images they yield are necessarily metaphorical. We don't ever "see" atoms. The images we obtain are indeed based on a stable, mind-independent reality. The predictive power and utility of the images derived from X ray and STM experiments are very impressive. One is moved to think, "Surely we are really seeing the atoms here!" What we see are constructs that at their best represent reliable models of reality, with sufficient verisimilitude to serve as productive metaphors. They facilitate correlations, predictions, and interpretations of other data and stimulate the creative design of new experiments. That is all we can hope for.

6

MOLECULAR MODELS IN CHEMISTRY AND BIOLOGY

On a wintry February 28, 1953, in Cambridge, England, James Watson and Francis Crick put into place the final pieces of a puzzle on which they had been laboring feverishly for months: the double helical structure of DNA. Watson recalled that Crick announced later that day to the patrons of the Eagle Pub that "we had found the secret of life." Well, if not the whole secret, they had certainly come upon a very important part of it. Their proposal for the structure of DNA, which has proven consistent in all its essential particulars with massive experimental evidences, ranks as one of the most significant scientific events of the twentieth century.[1]

Molecular models were one of the most important tools used by Watson and Crick in their effort to find a viable model for DNA. Figure 6.1 shows them posing with one of their final constructions. This and other homemade models facilitated the thought processes involved in formulating and testing various hypotheses regarding the structure of DNA. The models, based on general knowledge of the geometries of small molecules and limited X ray diffraction data on DNA itself, eventually led them to propose the famous double helix structure for DNA. Watson and Crick used reasoning based on analysis of three-dimensional objects in the macroscopic domain to understand the structure and properties of objects in the microscopic domain, beyond direct observation. In this chapter we explore the nature of mappings from models to molecules and some of the capabilities and limitations of molecular models as metaphors.

Figure 6.1. James Watson and Francis Crick pose with their model of
the DNA double helix. (Photograph by Barrington Brown; used by
permission of Photo Researchers, Inc.)

Origins

Molecules are formed by combinations of atoms. In the early days of chemistry it was discovered that elements combined with one another according to
certain rules. So, for example, carbon and oxygen were known to combine under certain experimental conditions to form carbon monoxide. The carbon
monoxide molecule is represented as CO, to denote that it contains one atom
of carbon and one of oxygen. Under different experimental conditions, carbon
and oxygen can form carbon dioxide, which is represented as CO_2 to denote that
there are two oxygen atoms present in the compound for each carbon. As more
and more data of this kind accumulated, the idea arose that the atoms of a giv-
en element have particular capacities for forming chemical links to other atoms.

This combining capacity was called valency. Mendeleev related the valency of an element to its position in his periodic table of the elements.

Great strides were being made in organic chemistry in the latter half of the nineteenth century. Chemists were learning how to synthesize new substances by using particular reagents, and many important practical advances were made in this period. But these practical advances occurred in the absence of a satisfactory theory of chemical bonding. Although chemists knew that the atoms of different elements were present in substances in certain ratios, they had no idea of how the atoms were connected or what the molecules formed by combinations of atoms might look like. Frederick Kekule, perhaps the greatest organic chemist of the period, represented the molecule of methane, CH_4, as shown in figure 6.2. This metaphorical representation was intended to convey the experimental fact that the methane molecule contains one carbon and four hydrogen atoms. Furthermore, it represents carbon as tetravalent. That is, Kekule assumed that the four hydrogen atoms were bonded individually to the carbon and not to each other.

Figure 6.2. Kekule's representation of the methane molecule.

Molecules must clearly be three-dimensional objects, but at the time there were none of the powerful tools for imaging molecules that were described in chapter 5, such as X ray diffraction. Kekule's representation of methane and the models used by other chemists of the time to represent molecules were two-dimensional in character. Such metaphors did not have the capacity to capture aspects of chemistry that had already been established from experiment. Some of the leading lights of chemistry, perhaps influenced by strongly anti-atomist sentiments prevalent at the time, thought that even to search for structural representations of molecules was futile. But it turned out that progress in understanding molecular structures could be made. It required imaginative deductions from indirect evidence and the use of well-chosen metaphors.

Molecular Asymmetry

In the early nineteenth century William Nicol, a professor at the University of Edinburgh in Scotland, gained immortality of a certain kind by inventing a special prism that bears his name. The Nicol prism has the property that when ordinary light is passed through it, plane-polarized light comes out. We can best appreciate the nature of plane-polarized light by thinking of light as a wave phenomenon. Suppose a beam of light is coming toward you, as illustrated in figure 6.3. In ordinary light, as it might come from the sun or a light bulb, the waves of radiant energy can move in any plane, as shown by the double-headed arrows in the figure. When the light has passed through a polarizer, such as a Nicol prism, only one of those planes has been selected, depending on the orientation of the prism. (The polarizing film used in sunglasses is also capable of selecting only one plane of polarization, thus reducing the glare from reflected light.)

The phenomenon of polarized light inspired many experiments. Jean-Baptiste Biot, a professor of physics at the Collège de France, studied what happened when plane-polarized light was passed through a sample tube containing a liquid substance or a solution, such as sugar in water. In most cases, nothing interesting occurred; the light passed through with no change in its plane of

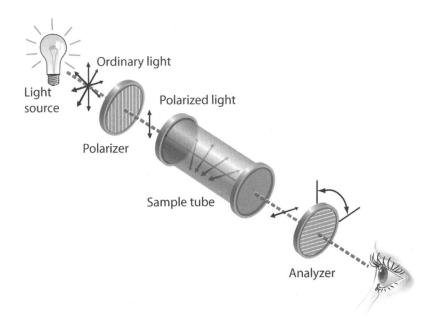

Figure 6.3. Schematic of the formation of plane-polarized light and its rotation by an optically active sample. The maximum brightness of transmitted light is attained when the plane of polarization of the light is aligned with the polarization of the analyzer as it is rotated. The instrument thus measures the angle of rotation of the plane of polarization.

polarization. For some samples, however, such as camphor or solutions of sugars, the plane of polarization was rotated. The light came out of the tube polarized in a plane that made an angle with the original plane of polarization, as illustrated in figure 6.3. Sometimes the rotation of the plane was clockwise, sometimes counterclockwise. Biot called the phenomenon optical activity. He conjectured that the molecules of optically active substances must be asymmetric in some way. But because there was no theory of how molecules are "constructed," no one could say what it meant for a molecule to be asymmetric. There was as yet no answer to the question of why some substances exhibited optical activity and others did not.

Early in his career, the French chemist Louis Pasteur had occasion to study certain salts of tartaric acid, an organic compound formed in the course of winemaking.[2] In looking at crystals of the compound under the microscope he discovered that they were asymmetric in shape. They were of two kinds, which were mirror images of one another, like the right and left hands. He painstakingly separated the two kinds of crystals with a pair of tweezers. A solution of one of the mirror-image crystal sets caused rotation of plane-polarized light in the clockwise direction. A similar solution of the other set of crystals rotated the light an equal amount in the counterclockwise direction.

At age 26, Pasteur was worried about making a blunder. Biot himself, by then a distinguished member of the French Academy of Sciences, had grown crystals of the same salt of tartaric acid and had failed to note the distinctive crystals Pasteur had found. Furthermore, when Biot and others had dissolved their crystals in water and checked the solutions for optical activity, none had been noted. Pasteur successfully repeated his experiments in Biot's presence, thereby convincing him that Pasteur had succeeded in separating two forms of an asymmetric molecule.[3] Pasteur's experiment was important because it showed that the phenomenon of optical activity could be associated with molecular properties that manifested themselves in the macroscopic domain, in this case in the shapes of crystals grown from solution.

In addition to the existence of optically active molecules, other observations of organic substances called out for an explanation. For example, it was known that two substances with closely similar chemical properties and the same molecular formula could have different physical properties, such as melting or boiling point. What made these substances, called isomers, different? In all likelihood, it had to do with some difference in the ways in which the atoms were connected or arranged in space. Kekule and others had developed a theory of organic structure that dealt with which atoms were connected to which. However, in the absence of experimental data on molecular structures, and with no theory of the spatial character of organic structures, only limited progress could be made in understanding optical activity and the existence of many types of isomerism.

Tetrahedral Carbon

The impasse was broken in 1874, when Jacobius H. van't Hoff and Joseph A. LeBel independently published papers that addressed the riddles posed by the work of Biot and Pasteur. Both authors came to roughly the same conclusions, but van't Hoff's paper was more graphic. In addition, he went beyond the discussion of optically active substances to address more general problems of isomerism. Van't Hoff therefore is generally credited as having a more influential role in the development of ideas about the three-dimensional structures of molecules.

At the heart of both papers was the idea that the explanation for optical activity lay in the arrangement of the groups attached to the carbon atom. Van't Hoff presented a list of optically active substances. He noted that according to Kekule's theory of structure, each substance in that list contained at least one carbon atom bonded to four different groups. How could this fact be connected with optical activity? Van't Hoff realized that if the four groups are arranged around the carbon such that they are at the corners of a tetrahedron, the resulting molecule must be asymmetric in a special way: It is not superimposable on its mirror image. Figure 6.4 depicts a molecule with four different groups attached to a central carbon and its mirror image. The molecule on the left cannot be superimposed on the one on the right, just as you cannot superimpose your left hand on your right. (The term *superimpose* in this context does not mean simply face-to-face matching, as when you place your left palm against

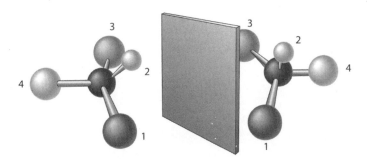

Figure 6.4. A molecule with four different groups attached to the central carbon and its mirror image, or enantiomer. The molecule on the right is not superimposable on the molecule on the left.

your right. It means that if one of the pair occupies the same space as the other, there will be a perfect match. If you imagine moving the molecule on the left in figure 6.4 to the molecule on the right, then rotating it in any way you choose, you will see that there is no way that all four of the groups on the central carbons can be made to superimpose on one another.) Molecules that are nonsuperimposable in this way are called enantiomers (from the Greek word *enantios,* a mirror). It is just this kind of asymmetric character that was needed to account for the existence of optical isomers.

Although the idea of a tetrahedral geometry for tetravalent carbon seems obvious to us today, in 1874 there was no theory of chemical bonding, only ideas about the capacity of an element to bond to other atoms. Nor were there any direct experimental ways to learn about molecular structures. Van't Hoff and LeBel offered a framework for interpreting many existing results and for making cogent predictions.[4] For the first time, molecules were conceived in specific three-dimensional terms. The impact made by van't Hoff and LeBel's papers rested on the particular metaphor that the four atoms attached to a central tetravalent carbon atom are disposed at the corners of a tetrahedron. It is important to notice that the four atoms attached to the carbon could have other atoms attached to them. For the purposes of creating the necessary asymmetry, it does not matter whether the four groups in figure 6.4 are single atoms such as H or Cl or are instead a group of atoms, such as CH_3. Whatever their particular identity, the two molecules related by the mirror-image symmetry are not superimposable. Van't Hoff and LeBel were able to account for the existence of optical activity in substances such as tartaric acid and many other known optically active substances.

Van't Hoff and LeBel's metaphor consisted in these elements:

> Molecules are spatially extended.
>
> a. They typically have specific three-dimensional structures, which remain fixed over time.
>
> b. The properties of a substance at both the molecular and macroscopic levels are related to its three-dimensional structure.

Let's examine the significance of each of these elements.

In his published paper, van't Hoff offers drawings of tetrahedra and other geometric figures as an integral part of his presentation. It is said that when he sent copies of his paper to other scientists he included cardboard models to aid in understanding the work. The model of a tetrahedral arrangement of bonds around carbon is obviously the product of experience with spatially extended objects in the macroscopic world. LeBel's paper is more formal in structure, but it nevertheless is based on imagery of macroscopic objects, ones that can be imagined either to have or to lack certain symmetry properties.

The tetrahedral arrangement around the carbon atom shown in figure 6.4 could be rearranged if the four groups attached to the carbon were able to move with respect to one another. However, if the properties of the molecule based on its asymmetry are to be experienced through laboratory observation, such as rotation of the plane of polarized light, the molecule must retain its asymmetric structure over the duration of the experiment. The persistence of optical activity means that the bonds around the carbon are sufficiently rigid so that such rearrangement does not occur. As it happens, there are cases in which the bonds are not rigid or some chemical process occurs causing the loss of asymmetry at the carbon. In such cases, the molecules do not appear to be asymmetric.

The general assumption that microscopic, atomic-, and molecular-level properties constitute the macroscopic properties of a substance underlies all applications of models. It is supported by such phenomena as that discovered by Pasteur: An asymmetry in the macroscopic crystals reflects the underlying asymmetry of the molecules that form the crystal. However, there is not always such an obvious connection between the molecular and macroscopic levels. For example, when Pasteur's tartaric acid salt is crystallized at higher temperatures than that which he used, the crystals do not form separate enantiomeric mirror-image forms. Rather, a symmetric crystal that contains equal numbers of the two enantiomeric molecules is formed. If Pasteur had not luckily come upon crystallization at the lower temperature, he would not have possessed the critical element in the observation domain on which to base his hypothesis of asymmetric molecules.

Van't Hoff and LeBel's models, derived from their direct physical experiences, were being applied to objects that were not observable at the molecular level. Today, scientists have powerful tools for probing the microscopic structure of matter. Yet molecular disciplines such as chemistry, biochemistry, and molecular biology rely as much as ever on the same kind of reasoning used by van't Hoff and LeBel, from the shapes and sizes of objects and images in the macroscopic domain to structures and processes at the molecular level. We will shortly see some examples of contemporary applications of molecular modeling.

Molecular Shape and Size

Van't Hoff and LeBel introduced the idea of geometry into chemistry. But although it was possible to think in a formal way about molecular shapes and symmetries, there was at that time no real theoretical rationale for the proposed structures. Why were the four groups bonded to carbon in methane and related molecules disposed tetrahedrally, instead of in some other way, such as in a plane around the carbon? Later, when the atom had been revealed to have a nuclear structure, with the electrons occupying most of the volume of the atom, it was reasonable to propose that the entities involved in forming connections between one atom and another were electrons. The earliest modern theories of chemical bonding modeled the chemical bond as a sharing of electrons between two atoms or, in the case of ionic compounds such as sodium chloride, a transfer of one or more electrons from one atom to another.

To pursue our inquiry into the metaphorical nature of molecular models, we need not delve into the details of chemical bond theory. For our purposes it is sufficient to focus on the ideas illustrated in the models for methane, CH_4, shown in figure 2.4 (p. 23). Recall that in the ball-and-stick representation for methane, the bond between carbon and one of the hydrogen atoms is modeled as a stiff rod. In modeling the methane molecule we need only focus on the fact that the bonds to the four hydrogen atoms are disposed about the carbon in a tetrahedral arrangement.

The model displayed in the lower part of figure 2.4 conveys other important aspects of the methane molecule. Recall from the discussion of earlier chapters that atoms are known to have spatial extent. When atoms or molecules approach one another they may experience weak attractive forces. However, on close collision a repulsive force dominates as the electron clouds of the non-bonded atoms begin to overlap. Assuming that the colliding entities are not chemically reactive toward one another, the collision is like that of a pair of soft billiard balls, as described in chapter 2. The apparent radii of the atoms in such collisions are known as van der Waals radii. The models of a few simple molecules are shown in figure 6.5, with the atoms shown as having van der Waals radii. Such models are called space-filling. It is helpful to remember that the volumes represented in these models are created by negatively charged electrons. The nuclei are buried in the center of each atom.

Before the advent of readily available computer programs, models such as those shown in figure 6.5 were created by assembling pieces that snap together, like the elements of a toy set. Today the models are more commonly created as computer-based images, using programs that generate the model from input information regarding which atoms are connected to which and by what kinds of bonds. The programs have achieved a great deal of sophistication and are capable of handling very large biological molecules as well as solid-state structures. The programs are also capable of accepting as input the structure of a molecule as determined from X ray diffraction or some other experimental method of structure determination.

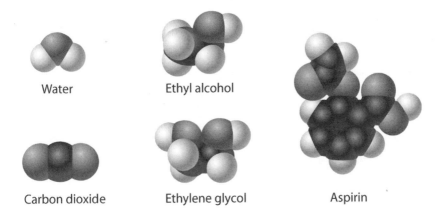

Water   Ethyl alcohol

Carbon dioxide   Ethylene glycol   Aspirin

Figure 6.5. Space-filling models of several small molecules.

Molecular modeling programs have become powerful research tools in material chemistry and molecular biology. We will examine a few selected examples of their uses, taken from the recent scientific literature, to learn how differing spatial metaphors for molecules are used, what aspect of the microscopic domain each purports to model, and what aspects are hidden. In so doing, we will see how strongly our understanding of the molecular domain rests on metaphors framed in terms of both structures and processes characteristic of macroscopic objects.

Molecular Cages

The April 25, 1997, issue of the journal *Science* contains an article titled "Nanoporous Molecular Sandwiches: Pillared Two-Dimensional Hydrogen-Bonded Networks with Adjustable Porosity."[5] Notice that the title uses expressions that map from the domain of everyday experience to the molecular domain: "molecular sandwich," "pillared . . . networks," "adjustable porosity." The article concerns the synthesis and characterization of a new class of solid-state materials that have the capacity to take up other substances within the solid material. The authors characterize the materials they report as "nanoporous lattices, which typically behave as hosts for molecular guests." The term *nanoporous* means that molecules of the sizes of those shown in figure 6.5 can penetrate into the solid. The word *lattice* is used to indicate that the solid has a regular repeating structure. The solid is characterized as behaving as a "host" for the smaller "guest" molecules that penetrate into it. Metaphorical mappings such as these, from the social domain to the microscopic domain, often are used to conceptualize actions or relationships between entities.

The intent of the work was to create new materials with void spaces that could be occupied by guest molecules. The synthesis of the new materials was planned on the basis of experience with substances that showed some of the desired properties, but in different contexts. The authors anticipated that their approach would lead to "the formation of voids in 2D [two dimensional] galleries with sizes, heights, shapes and chemical environments that could be manipulated by choice of molecular pillar." The use of terms such as *galleries* and *pillar*, in company with reliance on the three-dimensional models of the solid lattice, suggest reasoning about molecular-level structures on the basis of metaphors drawn from the macroscopic world.

The structure shown in figure 6.6 is a molecular model based on X ray diffraction studies that establish the spatial locations of the lattice framework atoms. The figure shows only a short section of the lattice, which extends back into the page and out of the page. Thus, the solid structure can be seen to consist of long, nearly rectangular channels. In this figure a typical guest molecule occupies the channels.

It was observed that the structure changes in response to the presence and particular nature of the guest molecules in the channels. A possible reason for this is that the guest undergoes both attractive and repulsive interactions with the atoms that make up the walls of the channel. As the guest molecule increases in size, it is necessarily in closer contact with the walls. Close contact with the walls turns a weak attractive interaction into a repulsive one. We can sense from figure 6.6 how this might happen.

Models such as those shown in figure 6.6 are very useful to the chemist. They convey a sense of the spatial relationships that are important in determining the properties of the system. For example, it would be possible to estimate from the models just how large a guest might be fitted into the channels. Also, the way in which the lattice tilts and expands or contracts, depending on the size and shape of the guest molecule, can be accounted for in terms of the spatial relationships revealed by the models. Useful as they are, however, such models are not literal depictions of the physical reality under study. Even with good X ray data, uncertainties remain regarding the precise locations of atoms in the lattice. In addition, all such molecular assemblies are continually in motion. Each atom vibrates around its normal position, so the entire lattice can be thought of as somewhat flexible. This structural flexibility generally increases with increasing temperature. Finally, as we have already learned, atoms are not hard spheres. Guest molecules encounter not static, hard surfaces, as implied by

Figure 6.6. A space-filling molecular model of a lattice and guest
molecules. (Reprinted with permission from V. A. Russell, C. C.
Evans, W. Li, and M. D. Ward, "Nanoporous Molecular Sandwiches:
Pillared Two-Dimensional Hydrogen-Bonded Networks with Adjustable Porosity," *Science* 276 [1997]: cover. © 1997 American Association for the Advancement of Science)

figure 6.6, but rather shifting, softer, more nebulous surroundings against which they move and from which they are reflected.

So the molecular models are incomplete and not entirely accurate, but does this make them metaphorical? A literal description of something can also be incomplete and inaccurate. The statement "That is a calico cat" is an incomplete description of a cat. Nevertheless, the statement can be entirely literal. Molecular models are metaphors because they represent a mapping from the domain of pictorial or three-dimensional model representation onto the domain of data from X ray diffraction and other experimental observations. The latter do not in themselves have anything three-dimensional or pictorial about them. For example, the model entails an image of small molecules moving easily through the "galleries" created by the arrangement of the molecules of the solid. This feature maps onto observations that small molecular substances are readily taken up and move rapidly through the solid. The model enables us to use concrete, familiar material to understand observations that are otherwise devoid of direct physical significance.

Protein Structure and Function

The many processes that continually occur in living cells, which determine the life of the cell, are controlled by proteins called enzymes. Enzymes are naturally occurring catalysts. They facilitate chemical transformations in living systems without being consumed in the processes. Over the course of evolutionary history, many enzymes with highly specialized functions have evolved. Typically, an enzyme catalyzes a specific kind of chemical process, and it operates only on specific molecules or on a class of related molecules. Enzymes often need the presence and participation of one or more other substances, called cofactors, that participate in the catalysis.

The understanding of how enzymes operate has been greatly facilitated by the use of X ray diffraction to determine their structures. Such studies are beset with many difficulties because enzymes are large molecules or clusters of molecules. For example, the enzyme fumarate reductase is formed as a cluster of four proteins, with a total molecular mass of 121,000 mass units. This is to be compared with the mass of 120 for the aspirin molecule shown in figure 6.5. Because enzymes are so large, their structures cannot be determined with the same degree of precision as the structures of smaller molecules. To help interpret the experimental data, scientists use molecular modeling programs designed to replicate some of the characteristic features of protein structures. The structural model that results from such procedures is grounded partly on the data for the enzyme itself, partly on models for related proteins, and partly on assumptions regarding particular aspects of the proteins that make up the enzyme.

In an article titled "Structure of the *Escherichia coli* Fumarate Reductase Respiratory Complex," T. M. Iverson and coworkers report their determination of the structure of an important protein.[6] As noted earlier, the enzyme consists of four subunits that come together to form the complete functioning enzyme. There are more than 8,400 atoms in the complete structure. In microorganisms the enzyme spans the cell membrane. Part of it projects into the solution outside the cell, and a part is within the cell. In the middle there is a membrane-spanning section.

Solving the structure of such a complex entity entails first of all obtaining it in crystalline form, which in itself is a daunting challenge. The structure obtained is thus not that of the enzyme as it exists in the cell but rather as it exists in a crystalline environment in which only the enzyme is present, along with smaller molecules, such as water. The hope is that the structure does not differ in an essential way from the structure of the active enzyme in the cell. Fumarate reductase was found to exist in the crystalline state as paired molecules. The space-filling model of the enzyme pair is shown in figure 6.7. The two parts of the pair are not strongly bonded to one another but are held in proximity by weak attractive interactions. The interest therefore centers on either one of the two molecules as representative of the enzyme.

Fumarate reductase catalyzes the transfer of electrons across the cell membrane to a particular chemical in the cell, the fumarate ion. The reaction is a means by which microorganisms obtain energy. In fumarate reductase, the electron transfer occurs in a series of steps, something like a ball bouncing down a stairs, where the steps represent energy changes. (Notice that once again we find ourselves using the location metaphor, in which changes in the state of the system are conceptualized as movements in space.) To understand the action of the enzyme, scientists must understand each step and how it relates to the others. Each step in the process involves one or more small molecules, called cofactors. To understand the overall process, the nature of the immediate chemical environment around each cofactor must be known.

The space-filling model shown in figure 6.7 is not suitable for examining the local structure around the sites in the enzyme where steps in the reaction occur. A ball-and-stick model would be better, but the structure would still be very crowded. A common practice in analyzing such complex systems is to represent the enzyme structure more symbolically. To see how this is accomplished, we begin with the fact that proteins are made up of amino acids. The amino acids are connected through chemical linkages, so the protein is a bit like a chain of

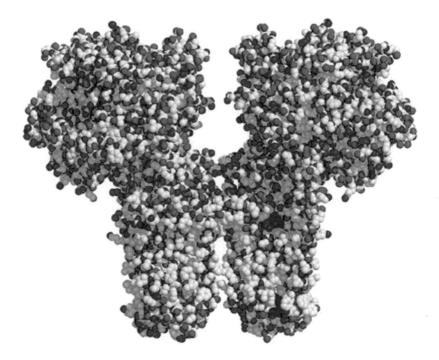

Figure 6.7. A space-filling model of the fumarate reductase molecular pair as observed in the crystalline state. (Reprinted with permission from T. M. Iverson, C. Luna-Chavez, G. Cecchini, and D. C. Rees, "Structure of the *Escherichia coli* Fumarate Reductase Respiratory Complex," *Science* 284 [1999]: 1962, fig. 1A. © 1999 American Association for the Advancement of Science)

beads. However, although all the beads in a chain might be alike, in proteins the "beads" can be any one of twenty different amino acids that form the basis of nearly all proteins. The order of the amino acids along the chain gives what is called the primary structure of proteins. Figure 6.8 shows this schematically as a random chain. (We will have more to say about this form of the protein in the next chapter.) Analysis of the structures of many proteins has revealed certain patterns in the ways the chain of amino acids is arranged in space, as illustrated schematically in figure 6.8. The shapes taken up by segments of the chain make up the secondary structure. In some sections of the chain the amino acids form a helical coil. In others they form a structure that resembles a pleated sheet, consisting of lengths of amino acids running alongside one another.[7] Regions between the helical and sheet sections consist of turns, loops, and less

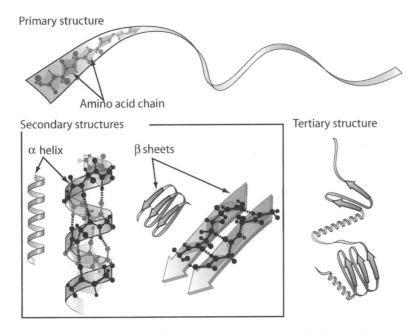

Figure 6.8. Schematic illustration of the various structural forms characteristic of proteins. The primary structure is determined by the linear sequence of amino acids. The secondary structure consists in chain segments that have an α helical or β sheet structure. The entire amino acid chain that forms the protein consists of connected segments of secondary structure. These come together to form the tertiary structure of the protein.

organized lengths. The entire chain forms a large, compact structure, composed of various helical or sheet sections and their connections. This compact assembly is called the tertiary structure. The protein in its most stable form, or "native state," is said to be "folded." (We encountered the metaphor of folding in chapter 2 and will examine it in more detail in chapter 7.)

Computer programs have been developed to examine a protein structure, such as that shown in figure 6.7, and identify the various secondary structures along the chain. Then the secondary structures are represented in a highly abstract form, similar to that shown in figure 6.8, as coiled or flat ribbons, with lines to denote bending regions or sections of chain that are not in a helical or sheet form. Figure 6.9 shows such a representation of fumarate reductase.

In figure 6.9, the details of individual atoms and chemical bonds shown in figure 6.7 have been discarded. The emphasis is on the general form of the

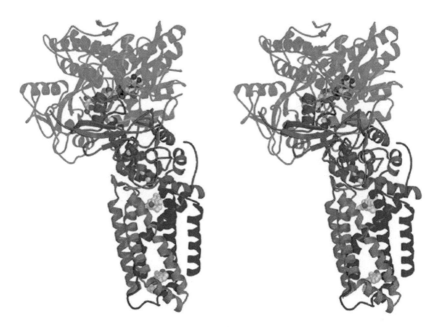

Figure 6.9. A stereo view of fumarate reductase, with protein sections shown in schematic form. If you hold the book at a close but comfortable reading distance and relax your focus, as though looking beyond the page, a 3-D representation should emerge in the center of your view. (Reprinted with permission from T. M. Iverson, C. Luna-Chavez, G. Cecchini, and D. C. Rees, "Structure of the *Escherichia coli* Fumarate Reductase Respiratory Complex," *Science* 284 [1999]: 1962, fig. 1B. © 1999 American Association for the Advancement of Science)

protein and on the dispositions of the various sections in space. The helical ribbons of course represent helical sections of the protein. The flat ribbon segments represent sections with the β sheet structure. An arrow embedded in the chain denotes the direction in which the amino acids point along the chain; we need not concern ourselves with this detail. The four subunits that make up the enzyme are shown. Each subunit is a continuous chain of amino acids, not bonded directly to the others. Notice also that the model shows groups of space-filling molecules dispersed within the structure. These represent the various cofactors involved in the electron transfer process facilitated by the enzyme.

What can we make of the multiple metaphors represented by this image? With all atomic-level detail of protein structure abstracted away, the protein is seen as a stylized structure, with regions of secondary structure connected by bends or unfolded lengths. We get a picture of the overall shape of the enzyme.

The relationships of the four subunits to one another are made clear as they could not possibly be in a more detailed rendering. Furthermore, the placement of the small molecule cofactors within the structure is made evident. In this metaphor, the protein sections take on a different identity: They are seen as a scaffolding that supports the various assemblies of cofactors involved in the electron transfer process. The important considerations are how the enzyme makes space for the cofactors and how the cofactors are attached to the protein. Such attachments have the function of keeping them in a particular location relative to other components. In addition, the chemical links that form the attachment may also activate the cofactors toward reaction.

To look more closely at the immediate surroundings of an active site, we need a close-up view. One such view used by Iverson and colleagues is shown in figure 6.10. Here we see still another set of metaphorical representations of the same observational data. The zigzag chains represent portions of the amino acid chains that comprise the various subunits; parts of three subunits occupy this region of the structure. The space-filling model in this figure represents an iron-sulfur complex that is one of the cofactors. This grouping of atoms is attached to the protein through bonds to particular amino acids, one of which is labeled in the figure as Cys B210. Another molecule is shown as a variant of the ball-and-stick model. This molecule is the source of the electrons transferred by the protein to the fumarate ion. The model calls attention to the overall shape of

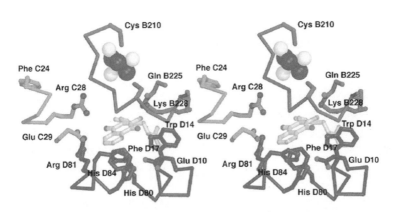

Figure 6.10. A stereo-view model of a portion of the fumarate reductase enzyme complex. (Reprinted with permission from T. M. Iverson, C. Luna-Chavez, G. Cecchini, and D. C. Rees, "Structure of the *Escherichia coli* Fumarate Reductase Respiratory Complex," *Science* 284 [1999]: 1963, fig. 4A. © 1999 American Association for the Advancement of Science)

the molecule and the manner in which it fits into a kind of pocket in the protein. We need not go into details of how the enzyme functions; the main point is that the form of display is chosen to highlight particular features, in this case the positioning of the cofactor in the enzyme and its relationship to the molecule that is the source of electrons.

The representations of the fumarate reductase structure discussed here are all based on the same data set and on the same underlying models and assumptions about what the data mean. Each model is a metaphor designed to highlight certain features of the enzyme structure. Inevitably, highlighting one set of features requires that other features are hidden. The space-filling model comes closest to portraying the nonbonded contacts between atoms. It also shows the overall outer shape of the enzyme and provides a basis for modeling how the enzyme might fit into the cell membrane.

The secondary and tertiary structures of the proteins are obscured in the space-filling model. To show these, the authors provide an inside view of the assembly in the form of the symbolic representation shown in figure 6.9. The stereo view further deepens the sense of how the components of the enzyme are arranged with respect to one another and how the cofactors are accommodated. Details of the local environments about cofactors and reacting molecules are shown in yet another schematic form, as shown in figure 6.10. Here, the structure has been abstracted in a different way to reveal the specific amino acids along the chains that are in close proximity to the cofactor and reactant. The models portrayed in figures 6.9 and 6.10 do not tell us anything about specific interatomic distances, nor do they give a clear picture of the nonbonded atom-atom contacts. Both figures are mixed metaphors in the sense that multiple metaphorical representations of atomic assemblies are used in the same figure.

The various views of the enzyme or parts of it all map onto the same set of data as processed via the theory of X ray diffraction. Each model calls attention to a distinct interpretation of the data, whether it be in terms of the space-filling character of the atoms that make up the protein, the presence of characteristic secondary structures, the relationships of the four subunits to one another, or the accommodations of cofactors within the structure. The metaphors enable us to understand the X ray data in terms of representations of three-dimensional, macroscopic scale objects. Multiple models are used to understand various aspects of the observation domain, just as multiple metaphors are used to understand an abstract domain such as argument (e.g., "An argument is a journey" or "An argument is a construction"). Each model hides certain properties or components of the system under study to convey more clearly a particular aspect. If the parts that are hidden have relevance to the highlighted aspect, the simplifications effected by the change in metaphor may lead to false conclusions. Whether a particular use of molecular models is appropriate is thus a matter of judgment and must be subjected to critical analysis in evaluating work that uses such metaphorical constructs.

What Are Models For?

The models encountered in this chapter are quite clearly mappings from structures and images in the macroscopic domain to experimental data on systems in the microscopic domain, beyond the range of direct observation. It is important to appreciate the deep nature of these metaphors. The models are macroscopic objects or computer simulations of objects that we can experience directly or via computer simulation. They entail volume elements that occupy mutually exclusive regions of space: joints in the structures that can be twisted in only certain ways to give shape to the structure, natural alignments of groupings of atoms in certain ways that seem natural in the structure, and so on. All these properties of macroscopic objects are mapped onto molecular-scale objects. We impart to microscopic objects properties analogous to those of macro-scale objects with which we have direct experience. These metaphors are so commonplace that it becomes very easy to mistake them for literal descriptions of the molecular scale. But they can't be literal descriptions. We do not have access to sensory data or data derived from instrumental extensions of our senses that reveal any of these aspects of molecular-scale objects in a direct sense. Rather, we have data that we can interpret in terms of metaphors in an "as if" sense. The system behaves as if it were a collection of certain atoms connected in a particular manner, as if it were folded into this or that shape.

The mappings from models to molecules are in general not direct; the data collected through observation can be related to the models only through metaphors for interpreting the data. For example, the interpretation of X ray data in terms of a three-dimensional array of atoms involves models for scattering of X rays by atoms, models for the nature of the crystals (e.g., the degree of crystalline order, motions of atoms in the lattice), and models for the relationship of the intensities of the scattered X rays to lattice structure. Because models of atomic-level structure are based on multiple assumptions and simplifications, they are subject to uncertainties that are generally not evident from examination of the models themselves.

But why is it of interest to know the structures of atomic and molecular systems? There is first of all the matter of simple curiosity and the urge to classify things. Aristotle, the first great classifier, believed that it led to understanding.

In the era of modern science, studies of structure have been driven by the idea that structures determine properties. In proposing tetrahedral bonds about carbon, van't Hoff aimed to account for optical activity, a property of matter observed in the laboratory, in terms of an atomic property. A knowledge of structure opens the possibility of understanding chemical behavior that depends on that structure. Thus, the authors of the article on fumarate reductase close with this sentence: "This gives a more complete view of the respiratory chain at the atomic level, and increases our understanding of one of the most fundamentally important processes of biological systems."[8]

In summary, we have seen in this chapter that molecular models, representations of the three-dimensional structures of molecules, have been very important in the advancement of chemistry and molecular biology. Van't Hoff's and LeBel's model for tetrahedral carbon was more than a heuristic device. It proved to be the key element in the development of the entire field of stereochemistry, the understanding of many properties of substances in terms of their three-dimensional structures. In more recent decades, models for interpreting X ray diffraction and other related data have been among the most powerful tools in the advancement of modern chemistry and biology.

The key role played by models as essential elements in discovery is nowhere more dramatically illustrated than in the discovery of the structure of DNA. Watson and Crick based their double helix model of DNA on several experimental facts. They knew that the X ray data available to them could be interpreted to mean that DNA has a helical structure, and they could estimate the pitch, that is, how far apart each turn of the helix is from the equivalent point above and below it. Their model of the backbone of each component of the double helix was based on data for known structures of smaller molecular assemblies. They knew also that DNA is made up from four bases, labeled A, T, G, and C; that A and T were always present in equal amounts in any DNA specimen; and that G and C were also always present in equal amounts. The double helix model could account for this chemical fact in terms of bonding interactions between the bases on one strand with those on the other. It made good chemical sense that a base A in one helix must always be opposite a base T in the other; a base G must always be opposite a base C. The two strands are thus complementary.

The Watson-Crick model accounted for the existing experimental facts. Beyond this, however, it opened an enormous window on the molecular basis of life. The double helix model for DNA consists of two complementary strands of DNA. If the two helices are unwound, the cell has two equivalent strands, each of which contains all the information needed to reproduce the cell. Each strand, with the aid of suitable molecular machinery in the cell, builds a complementary strand onto itself, forming a new double helix. The Watson-Crick model thus went far beyond merely accounting for the data at hand. Francis Crick's claim at the Eagle Pub was quite understandable. He and Watson had indeed found one of life's most important secrets, with the aid of a homemade molecular model.

7

PROTEIN FOLDING

In previous chapters we have been concerned with metaphors that relate to structures. We've examined metaphors for the structures of simple entities such as atoms and small molecules and for more complex things such as extended inorganic solids and enzyme complexes. But why study structures at all? The answer is that our everyday experiences tell us that structures are related to properties. That a ball rolls readily on a surface is accounted for by its spherical shape. In the same way, as we saw chapter 6, a key property of certain substances, that their solutions cause rotation of the plane of polarized light, is accounted for by postulating the presence of four different groups arranged tetrahedrally around a central carbon atom. Because structure is so deeply entrenched in embodied reasoning, the idea of structure at the level of atoms and molecules arose even in times when there were no direct means of probing the spatial character of such small entities.

The relationship between structure and function is very important in many areas of science. By function we mean, What is the thing for? What role does it play in events that involve it? From a model for the structure of a thing, we may be able to deduce how it must function. But it works both ways: If we know how a thing functions, we can sometimes infer what structure it must have. The mapping between structure and function arises regularly in our everyday experience. Suppose you had never seen a vegetable peeler. If I present you with one and ask you to guess what its function might be, chances are you would be able to come up with something close to an optimal answer, especially if we were doing the exercise in the kitchen. In an analogous vein, the motivation for determining the structure of fumarate reductase, described in chapter 6, was that possession of a reliable model for its structure could lead to deeper insights into how the enzyme works.

In this chapter we will examine protein folding, an important process and a hot topic in current research. Two aspects of protein folding are important for our purposes. First, folding has to do with the relationship of structure to function. To carry out the functions we ascribe to them, proteins must adopt particular structures. When a protein lacks the correct structure, it is inactive with respect to the property for which it seems to have been designed. Second, folding is a dynamic process, involving change over time. We will see that the metaphors relating to processes, changes occurring over time, differ in important ways from those that apply to static structures. They rely on a different set of mappings from our everyday experiences in the macroscopic world.

As you read this chapter you may feel that you are being subjected to textbook stuff, but I'm not trying to make you into a protein chemist! In approaching the topic as I have, I hope to illustrate the importance of metaphors in learning new science. Every active scientist is a perennial student, regularly acquiring new ideas through reading and listening to others. The path we will take in building our understanding of protein folding is basically the same that any scientist new to the subject would take. It consists in seeing what observations have been made and the interpretations placed on them. We then look for connections with other observations and experimental work to situate the subject in a larger context. All of this forms the background for the development of metaphorical models that are both explanatory and consistent with larger, more general understandings of how nature behaves.

Introduction to Protein Folding

The term *protein* was coined by Jöns Jacob Berzelius in 1838. It is derived from the Greek *proteios,* which means "primary." Without much direct evidence to support it, Berzelius hazarded the guess that these newly discovered substances would prove to be the most important class of biological substances. Although the notion of a single most important class of substances in living systems is not plausible, Berzelius guessed correctly that proteins are very important. Each cell in a mammalian organism contains more than 50,000 different varieties of protein. Many function as enzymes, which serve as catalysts; that is, they facilitate, or speed up, biological reactions. We met an example of such an enzyme, fumarate reductase, in chapter 6. Other proteins are used for storage and transport: In the lungs hemoglobin binds oxygen in red blood cells and carries it to cells, where it is needed for cell activity. The protein insulin functions as a hormone, maintaining optimal levels of glucose; it signals the body when to convert excess glucose to a stored form. Still other proteins, such as collagen, provide structural support.

In the 1950s, Christian B. Anfinsen and his colleagues at the National Institutes of Health were studying an enzyme called ribonuclease, which consists of a string of 124 amino acids in a single chain. As a class, ribonucleases facilitate certain reactions in the cell. Like nearly all enzymes, ribonuclease must be in a particular condition, called the native state, to carry out the function for which it is designed. A change in environmental conditions or chemical treatment of a certain kind can cause a protein to become inactive. The inactive protein is said to be denatured.

It usually does not take a lot of heating or large changes in other environmental conditions to cause denaturation, which can be reversible or irreversible. Boiling an egg denatures albumin, the protein that makes up most of the white of an egg. Once denatured in this way, the albumin cannot be returned to its original active state by cooling the egg back to room temperature. The denaturation in that case is irreversible. However, Anfinsen and his colleagues found that ribonuclease can be denatured in such a way that, in the appropriate environmental conditions, the protein reverts to its native state. In the denatured state ribonuclease is inactive. Upon reverting to the native state, it regains its biological activity. The transformation from an inactive state to the native state is called folding. An analogous process converts proteins newly synthesized in the cell to their active forms, their native states. We will explore the nature of the folding process in detail as we go along. For now, picture the folded form of the protein as a compact structure, such as shown in figure 6.8 (p. 115) or 6.9 (p. 116). The unfolded state can be thought of as a long, random chain of amino acids.

Anfinsen showed that ribonuclease folds to the native state unaided by any extraneous biological molecules.[1] His seminal contribution lay in showing that a unique, biologically active, three-dimensional structure of a protein can be reached from the unfolded, denatured state without participation from other than the small molecules and ions in the solution surrounding the protein. Under appropriate conditions of temperature, degree of acidity in the solution, and other environmental factors, the protein reliably returns to a particular structure that depends only on the primary structure, the sequence of amino acids in the chain.

There are two important metaphors in what has just been related. The first is use of the term *folding* to characterize the process in which the protein changes from the unfolded, denatured state to the native state. As noted in chapter 2, the metaphor evokes the notion of bringing into contact various parts of the object, as in folding clothing or card table chairs. Folding is regarded as an appropriate metaphor for the process undergone by proteins because structural studies show that parts of the protein are brought into close contact to make a more compact globular entity from an initially extended chain. (Note the compact arrangement of the parts of fumarate reductase, as revealed in figure 6.9 [p. 116].) In some instances, such as folding a towel, the folding is not a uniquely determined process. In others, such as folding of a card table chair or a roadmap, the correct folding pathway is uniquely defined by the construction. Sometimes the proper pathway to the correctly folded structure is not so clear. You may have had the frustrating experience of trying to return a roadmap back to its original folded state. One often ends up with a jumbled approximation that lacks a key property of the properly folded state: the ability to pack neatly into the car's glove compartment. We'll see that the potential for frustration is also a factor in protein folding. In summary, the experiential gestalt associated with folding, with its various entailments, can be mapped onto the varied experimental observations regarding a characteristic process undergone by many proteins, called "folding."

The second metaphor involved in our discussion is embedded in the idea that the identity and order of the amino acids in the protein chain determines a unique or nearly unique final three-dimensional structure for the native state. This metaphor is commonly expressed by the statement that all of the information needed to determine the three-dimensional native state structure is contained in the linear sequence of amino acids.

Anfinsen showed that ribonuclease always folds to the same native state structure, but that fact alone does not mean that the "information" encoded in the amino acid sequence is responsible. Perhaps some other arbitrarily different amino acid sequence of approximately the same length would also fold into the same native state configuration. Anfinsen addressed this issue by making changes in the amino acid sequence. He found that many deletions of amino acids from the chain caused misfolding. Some other changes, such as substitution of one amino acid for another, had large effects; others did not. He concluded, "The correct and unique translation of the genetic message for a particular protein backbone is no longer possible when the linear information has been tampered with by deletion of amino acid residues. As with most rules, however, this one is susceptible to many exceptions."[2]

The metaphor that relates structure and properties in this case is taken from our understanding of language: "The order and identity of the amino acids in a protein is a language." We know that the meanings of written statements depend on unique or nearly unique linear sequences of letters and words. Sometimes a change of only a single letter changes the meaning or import of a statement. Compare two statements with the same pattern of letters and spaces: "The cat ran over the boy" and "The car ran over the boy," or "He shut the door" and

"He shot the door." Analogously, the devastating genetic disorder called sickle cell anemia results from a single amino acid change at position number 6 in the 146-amino acid protein beta-globulin.[3]

In other situations, changes that thoroughly mess up the sentence may nevertheless leave the meaning largely intact. The significance of the primary structure of proteins (the identity and order of amino acids along the chain) to the structure and function of proteins parallels the relationship of syntax (orderly or systematic arrangement) to semantics (meaning) in language. The gestalt that consists of the complex of associations and ideas that make up our understanding and use of written language maps onto the molecular domain of protein sequence. Notice that this mapping does not involve a directly emergent physical experience but rather a human artifact in the social domain. This is an early example of an important and interesting aspect of metaphors in science: As the scientist attempts to understand systems of increasing complexity, metaphors based solely on embodied physical experiences no longer suffice. In more complex systems, typically there are interactions between the components of the system. To deal with these interactions, which can be thought of as "transactions" between components, the appropriate mappings increasingly derive from social constructs, with their attendant greater complexities.

Some Properties of Proteins

Protein folding is one of the great unsolved problems in molecular biology.[4] Because it involves a change in the overall shape and internal relationships in a large molecule, it is a complex process. Understanding the process from a scientific point of view requires that many different aspects of it make sense in relationship to one another. We will touch on many of these aspects as we move along.

To appreciate more clearly what is involved in folding, we need to examine protein structure in further detail. Figure 7.1 shows a model of a short section of protein chain. The chain is made up of amino acid units; three are shown in figure 7.1. Ribonuclease has 124 amino acids; other proteins have several hundred or more. The different amino acids that occur naturally in proteins have the same structure except for the side group, attached to C_α, shown in figure 7.1 as large spheres. There are twenty-two different amino acids and thus twenty-two different side groups found in proteins, although not all twenty-two are necessarily present in any given protein.

The shaded areas along the chain represent the fixed part of all amino acids. The atoms of this grouping remain fixed in a plane, called the amide plane. The conformation, or arrangement of the chain in space, is determined by rotations of these planar groups with respect to one another, around the C_α-C

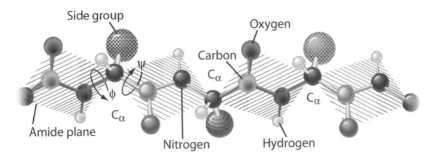

Figure 7.1. A section of amino acid chain.

and C_α-N bonds up and down the chain. These rotations are labeled as Ψ and ϕ in figure 7.1. By twisting about these bonds, the angles between the planes are varied, and the chain takes on various shapes. It is rotations about the C_α-C and C_α-N bonds that permit the amino acid chain to adopt the ordered α helix and β sheet structures described in chapter 6.

There are limits to the values that Ψ or ϕ can adopt in any given location along the chain. Rotations may cause the side groups on adjacent amino acids to bump into one another, depending on their shapes and sizes. Clearly, no two atoms or groups of atoms can occupy the same space. Various other constraints further limit the values that Ψ or ϕ can assume. As a rough general rule, there are three favored values for each Ψ or ϕ, although certain of these are ruled out by interferences between groups.

In addition to differing in size and shape, the side groups also vary in another important respect. Some are polar; this means that they interact in an attractive or favorable way with the aqueous environment in which the protein is immersed. Such groups are called hydrophilic, or "water-loving." Others are nonpolar, more like vegetable oil; they do not interact favorably with water. These hydrophobic, or "water-hating" groups prefer to stay out of the polar water environment.

Proteins in nature have evolved through natural selection to be highly effective in carrying out certain functions. To play on an earlier metaphor, nature has worked out a language based on amino acid sequences. A protein with just any random sequence of amino acids will almost certainly be read as gibberish; the sequence will not have a meaning expressible in terms of a function for the protein. An enzyme that catalyzes a particular chemical reaction may be exquisitely selective for only particular molecules, called substrates, on which it acts, or for only particular chemical bonds. The selectivity comes from the shape of the native state structure and changes in shape induced by interaction with the substrate. The enzyme is effective only if it is in the active, native state. Natural selection has yielded particular sequences of amino acids that, upon synthesis in the cell, readily form an active native state of the necessary selectivity and catalytic effectiveness with the target substrates. So, in summary, the amino acid sequence determines whether the protein folds and what shape it folds into; the folded structure determines the enzyme's activity as a biological catalyst.

But the protein's transition from a nonnative state to the active native state entails changes in energy. Recall that proteins in living systems exist in particular environments, mostly water that contains various dissolved ions (e.g., sodium, potassium, magnesium), and other molecular substances such as sugars, acids, and bases. It has interactions with all these components of its surroundings, and those interactions affect the energy of the system. Changes in the conformation of the protein (i.e., in the shape or arrangement of its parts) are accompanied by changes in energy. To understand the metaphors used in reasoning about protein folding, we must understand the metaphors used in reasoning about temperature and energy.

Temperature and Energy

We have all had the experience of being told, even as adults, "Don't touch that, it's hot!" Living organisms are quite sensitive to temperature variations, unable to survive in physical circumstances that fall outside a narrow range. We learn early in life that hot things burn us and that excessive cold causes us harm as well. The concept of temperature is directly emergent from these everyday experiences. In addition, we become aware of our own bodily temperature and the existence of homeostatic control. Even a slight increase or decrease in body temperature from normal is noticeable.

One of our very common experiences in connection with temperature is that when hot and cold things are placed in contact, the hot things lose their hotness, and the cool things warm up. For example, when we place an ice pack on a sprained ankle, our ankle is made cooler. The concept of heat derives from the idea that when two objects initially at different temperatures are placed in contact, something "flows" between them. Heat is a metaphorical construct; there is no observable entity, fluid or otherwise, observed in the process by which two bodies come to a common temperature. The observed processes of heat exchange occurring in nature sometimes are conceptualized in terms of directly emergent experiences with the flow of a fluid such as water. In the early scientific literature, heat, called caloric, was thought to be a weightless fluid. The metaphor of heat as a fluid with flow properties analogous to water was used by the French physicist Sadi Carnot in his pioneering treatise on heat, published in 1824,[5] Carnot extended the analogy to explain how work may be accomplished by the flow of heat from a higher to lower temperature. He conceptualized heat flow that produces work as analogous to the fall of water from a higher to lower level through a waterwheel.

Energy can be defined as the capacity to do work or to transfer heat. In a steam engine, hot steam is cooled, and in the process the heat that flows from it is converted, at least partially, into work. The chemical energy stored in an ordinary alkaline battery is also a form of energy. It is converted into work when it turns the machinery of a portable CD player, or it is converted into other forms of energy, such as radiant energy, or heat, when it is used in a flashlight.

Interconversions of heat, work, and mechanical and radiant energy are common in modern society, but the concept of energy as such is not directly emergent from our experiences. Instead, our ideas of energy derive from, and are understood in terms of, direct experiences with energy *changes* in various contexts. For example, the chemical potential energy stored in gasoline is converted in an automobile engine into heat and kinetic energy. In science, the focus is similarly on energy exchange and energy change.

Scientists conceptualize energy changes in terms of an orientational metaphor. Energy is represented by a vertical scale, using the "More energy is up" basic metaphor. This representation follows from our everyday experiences in Earth's gravitational field. The gestalt formed from experiences with falling objects, the effort needed to climb stairs, and so on is the source of the convention that less energy is down, more energy is up. For example, the water flowing through the turbines of Hoover Dam is moving to a lower vertical level and in the process losing potential energy. The energy lost is partially converted into the mechanical energy of rotating turbine blades and thence into electrical energy.

Energy Surfaces

Many processes that occur in nature involve a continuous change from one state of affairs to another. For example, in a chemical reaction, the molecules involved, called reactants, come together, new bonds are formed, and old bonds are broken. Chemists picture a process that leads continuously from reactants to the newly formed molecules, called reaction products. In such a process, energy changes occur. Often, for the reactant molecules to get into the correct relative position for reaction, they must be forced together. Energy is needed to do this, so the energy of the reacting system must increase. A commonly used metaphor for the chemical reaction is shown in figure 7.2. The event depicted

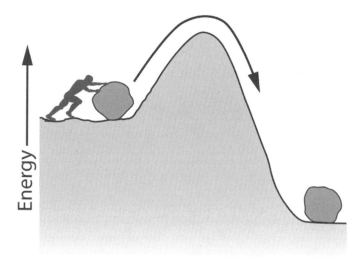

Figure 7.2. A schematic illustration of an energy surface in the macroscopic domain.

is movement of the boulder over a surface, from the left side of the hill to the opposite side. To accomplish the desired change, the boulder must be rolled up the hill, that is, taken to higher energy. Once at the top, it spontaneously rolls down the other side to come to rest at a lower energy than before.

Analogous diagrams are used in discussing the energy changes that occur in a chemical process. In figure 7.3 the boulder of figure 7.2 has been replaced by the assembly of atoms undergoing chemical reaction. The higher the position of this assembly on the energy surface, the higher the energy contained within it. When the peak is reached, the assembly spontaneously moves down the opposite slope to come to rest as product molecules. As before, energy is visualized as a surface. Just as in figure 7.2, the system—the portion of the world that is undergoing the change in state—needs a source of energy to overcome the barrier. The Sisyphus-like character in figure 7.2 is absent from figure 7.3; the means by which the molecule gains energy is not specified. In the underlying model, based on the kinetic theory of gases discussed in chapter 4, molecules exchange energy through collisions with other molecules.

The horizontal dimension in figure 7.3 is not a simple distance coordinate. It represents the progress of the reaction, which reflects many changes in the distances separating atoms within the molecules reacting, along with bond angle changes. The reaction process itself does happen to entail bond distance and angle changes; old bonds are broken, new ones are made, and molecular shapes

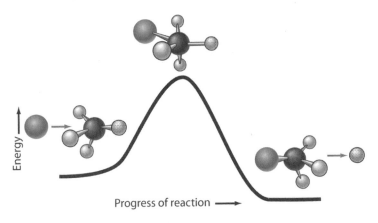

Figure 7.3. A schematic chemical reaction surface. The reaction is visualized as movement along the surface, with changes in energy that correspond to changes in the configurations of the atoms.

change. But progress of reaction appears as a length in figure 7.3 for a different reason. We see here another use of the location metaphor, as described in chapter 3. The change in state from reactants to products is conveyed metaphorically as movement from one location to another along a path that is specific for that particular change in state. This is a very general metaphor; we don't need to know the details of changes in bond angles and distances to use "reaction progress" as the horizontal dimension in the figure.

The energy of a complex molecule, such as a protein, depends on a very large number of bond distances and angles. Even when it is possible to formulate a mathematical relationship between energy and all these variables, there is no way to portray the relationship in a multidimensional graph. Each bond angle and bond distance would have to be a dimension, whereas our capacities for visualization are limited by our bodily experiences to three-dimensional space. But two-dimensional surfaces can still be useful. The many variables represented by all the bond distances and angles must be lumped together, in terms that mean something like "progress of reaction" or "progress of folding," just as was done in the one-dimensional example shown in figure 7.3.

As we move forward in our discussion we will be talking about models for protein folding. This is an appropriate point at which to pause, to remind ourselves of the relationship between the model, which is metaphorical, and the observation domain onto which it supposedly maps. What are the observations that provide us information on protein folding? We don't see long strands of amino acids change from a random chain into a compact three-dimensional structure. Instead we have the results of measurements made on the protein solution, from which we infer via our model that the change from random chain to a compact three-dimensional structure has occurred. The simplest observations are of the sort made by Anfinsen: The denatured protein is not functional as an enzyme, the native state, or "folded," form is. But "folding" refers to the process itself; how long does it take, and what changes occur in the protein along the way from the initial unfolded state to the final, native state? To explore these questions, a variety of experimental techniques are used. Many of them, so-called spectroscopic methods, involve use of light. The protein interacts with radiant energy differently as it undergoes the folding process. By monitoring the ways in which various kinds of spectroscopic signals change with time, the progress of folding can be inferred. Using such techniques, the data from laboratory studies are interpreted to mean that the first and fastest steps in folding are the assumption of sections of secondary structure, α helices and β sheets, followed by the rearrangements of these sections. More recently, fast X ray techniques provide data that can be interpreted in terms of changes in the protein's overall shape and size during the folding process. These studies also suggest that a rather compact structure forms early, then undergoes rearrangements to form the final native state structure.

The models we are about to discuss are metaphorical descriptions that, to be useful, must connect with the results of laboratory studies of protein solutions. Models for protein folding must have two basic components. The first is structural: The linear chain of amino acids must be represented in some way. Second, there must be a component that deals with the folding process itself: Which parts of the protein fold first and at what rates? The models for this aspect of folding are distinct from the structural models.

A Simple Model for Folding

Protein folding is akin to a chemical reaction in that there is a progression from one state of a molecular system to another. The large, complex amino acid chain changes from a randomly oriented string to an ordered structure. Scientists want to know why proteins change in this way. What is the driving force for the process; that is, what makes it go? To see what kinds of metaphors are useful in thinking about protein folding, we begin with a very simple model. One of the most important ideas in models of protein folding is that the hydrophobic (water-hating) side groups along the amino acid chain are clustered as much as possible inside the folded protein, away from the water environment in which the protein is immersed.[6] To see the basis of this idea on the macroscopic level, imagine putting a small dollop of vegetable oil in a container half-filled with water. The oil floats on the water as a single blob, illustrated in the top-down view shown in figure 7.4(a). Now, agitate the water and oil mixture vigorously, say with a blender. After the agitation, the oil blob is broken into a myriad of smaller, variously sized droplets, as illustrated in figure 7.4(b). If we let our little system stand for a time, we find that the many small droplets have coalesced back into the original single large blob. Why has this happened?

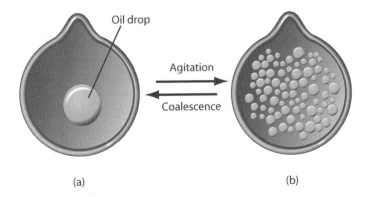

Figure 7.4. Illustration of the spontaneous coalescence of smaller oil droplets into a single larger one.

The molecules that make up the vegetable oil are nonpolar. The attractive forces between them are weak but sufficient to keep them together, forming a viscous liquid at room temperature. The vegetable oil molecules do not interact well with the polar molecules of water. For this reason, oil does not dissolve appreciably in water. The oil is in its lowest energy (most stable) state when the number of oil molecules that must interface with water is minimized. When the oil-water mixture is agitated and many small droplets are forcibly formed, the number of oil molecules in contact with water molecules is greater; the surface area of many small droplets is greater than that of a single large drop. Over time, the small droplets coalesce to form larger droplets. They do so because the surface area is thereby reduced. As a result, the system "moves to lower energy." It reaches it lowest-energy, most stable state when there is again just one large drop. The energy change involved in this process of coalescence is very small, but the process nevertheless occurs.

To see how this principle applies to protein folding, we'll consider a more directly interesting but still simple metaphor, illustrated in figure 7.5. In this case, the protein is conceptualized as a chain of beads.[7] Each bead corresponds to an amino acid. Instead of having up to twenty-two different amino acids, our chain has only two kinds of beads, hydrophilic and hydrophobic. The hydrophilic beads represent amino acids that have polar, water-loving side groups. The side groups on the amino acids represented by the hydrophobic beads are nonpolar, oil-like, water-repellent. To denote the various possible conformations of the protein, we imagine that our chain of amino acids is laid out on a grid, with each amino acid occupying a site in the planar lattice. That is, if the plane is ruled off in equally spaced vertical and horizontal lines, the chain is laid out along the lines, and the amino acids are situated at the intersections. The rules are that the chain of amino acids must stay on the lines (no diagonal crossings), and the chain must not cross itself. As you can see from figure 7.5, there are three possible angular arrangements at each amino acid (except for the two end ones), subject to the rules that no two acids can be at the same site, and the chain can't cross itself. When two hydrophobic groups that are not immediately adjacent along the chain occupy adjacent lattice spaces, that counts as a stabilizing interaction, shown as a dotted line in figure 7.5.

The lattice model is useful because it can be studied via computer programs.[8] Suppose we start from any typical unfolded configuration of the protein, such as the one shown in figure 7.5, and program a computer to search for all the possible configurations. There are a lot of them—on the order of a billion for the chain that is twenty-one amino acids long. By searching through all possible configurations of the chain, states that have the maximum number of stabi-

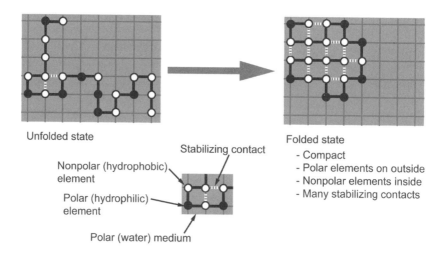

Unfolded state

Stabilizing contact

Nonpolar (hydrophobic)
element

Polar (hydrophilic)
element

Polar (water) medium

Folded state
- Compact
- Polar elements on outside
- Nonpolar elements inside
- Many stabilizing contacts

Figure 7.5. A lattice model for protein folding. Each amino acid on the protein chain is represented as a point on a square lattice. The amino acid residues are either hydrophobic (light) or hydrophilic (dark) in character.

lizing contacts between hydrophobic groups can be found. One such "folded" state is shown in figure 7.5. Notice that the folded state has a compact arrangement, with the nonpolar elements inside the chain and the polar elements on the outside. You can see that there is a similarity between this model and the one represented in figure 7.4. The hydrophobic groups move together to take advantage of their attractive interactions, minimizing their contacts with the external solution.

Now we come to a couple of interesting and subtle aspects of folding, a set of "folding meets statistics" ideas. First, we need to recognize that there are very few optimally folded forms in comparison with all the other possible configurations of the chain. Of the billion or so possible configurations of the twenty-one–amino acid chain in figure 7.5, only a couple of dozen have the maximum number of stabilizing contacts. Given this fact, the folded forms are extremely improbable. Should we expect to find our protein in one of them? Then there is the issue of how rapidly folding can be expected to occur. How does the protein get to a folded form, starting from any random unfolded state?

Let's deal with the probability issue first. Using the simple model illustrated in figure 7.5, it is possible to find and count the optimally folded structures, as measured by the number of stabilizing interactions between the hydrophobic groups. The number of fully folded forms depends on the particular mix and order of polar and nonpolar groups along the chain, but as already noted, it is tiny in comparison with the total number of possible configurations; most are unfolded or only partially folded.

To puzzle out the nature of the system, let's do one of those gedankenexperiments of the sort that Einstein liked to do. Suppose that we have a solution containing large numbers of molecules with twenty-one amino acids in the chain, with the mix of hydrophobic and hydrophilic side groups shown in figure 7.5. Let us further suppose that at some initial time we have somehow managed to have all the protein chains in the folded form shown in the figure. Finally, let's assume that there is no stabilizing interaction between the hydrophobic groups, so the folded form is no more stable than any other. Now we let the clock run and watch what happens.

We'll assume that our thought experiment system is at the temperature of a typical living system, say about 97° Fahrenheit, or 36° Celsius. Just as in a sample of gas atoms or molecules at such a temperature, the molecules of the solution are in constant, chaotic motion, jostling one another in the crowded environment of the solution. There are therefore very frequent exchanges of energy between the water molecules and the protein chains. The energy in the protein chains gives rise to twists in the bonds along the chain. There is therefore a continuous twisting, raveling, and unraveling of the chain so that any particular protein chain could at any given instant have any of the very large number of possible configurations for the chain. Although at the start of our thought experiment, all the protein chains are in the folded form, they would quickly evolve into a chaotic mixture of many different shapes. At the start of our experiment the system was ordered, in the sense that all the chains were folded. But remember that we are assuming for the moment that there is no energetic bias favoring the folded forms. After a time, there are very few if any chains left in the folded state because each folded state is so unlikely in comparison with the vast number of other, disordered states.[9] The system has spontaneously become disordered. It has obeyed the dictum of the nineteenth-century physicist Rudolf Clausius that although the energy of the universe is constant, the entropy, or disorder, of the universe tends spontaneously to increase. So we could say that chance favors the unfolded forms; there are just so many more of them.

But now suppose that while we have a general chaotic mixture of conformations present in our thought experiment, we "turn on" the stabilizing effect of near-neighbor contacts between the hydrophobic groups. Now we no longer have a level playing field. The folded states are now favored over the unfolded states. The larger the number of stabilizing contacts, the greater the degree of favoring. As the chains flop around and through happenstance come into a configuration where there are stabilizing interactions between hydrophobic amino acids, the chain will remain in that form for a longer time. The stabilizing energies oppose the tendency for the chain to take on just any configuration. If the stabilizing interactions are large enough, they can eventually lead the system to a state in which all, or nearly all, the chains are in the folded form, the one shown in figure 7.5 or another equally stable folded form.

This simple thought experiment illustrates a principle that is applicable to all systems in nature. The thermal energy of a system causes it to drift toward a random state and away from a single state that we might choose to call ordered. Thus, in natural systems there is a tendency toward disorder, toward increasing entropy. Opposing this tendency toward randomness, there may be a driving force that moves the system toward one state or set of states of lower potential energy. Just as a stone drops from the hand to the ground when released, because it moves to a lower potential energy, a molecular system also has a tendency to move from a higher to a lower potential energy. The net of these two opposing tendencies is a quantity called free energy. A system spontaneously moves to a state in which it is in equilibrium with its surroundings. This state corresponds to a minimum free energy.

We've covered a lot of new ideas in this discussion, but we need them to appreciate the nature of the metaphors used to model protein folding. To summarize: Proteins are driven to adopt characteristic folded states because of the attractive interactions that are present in those states. In our simple model we've assumed that the chief attractive interaction is the clustering of hydrophobic groups together inside the folded form, but other terms also contribute. Proteins are driven to *unfold* by the tendency of each molecule to adopt one of the many unfolded forms, which we characterize as disordered. Where a protein solution ends up when equilibrium has been established between the protein and its surroundings depends on the net result of these two opposing tendencies. No matter how sophisticated the model for protein folding, these principles operate just as importantly as in the simple model we have been considering.

Pathways for Folding

So where are we? We have seen that proteins typically are arranged in space in a particular way. Proteins that serve as enzymes, facilitators of the cell's chemical processes, are most commonly found in a compact and globular form. The change to the functional form from the long, random chain that characterizes the newly formed protein is called protein folding. We have also seen that a major driving force for this process is that the hydrophobic amino acid side groups are clustered in the folded form on the inside of the structure, out of the polar aqueous environment. The polar side groups are most often found on the outside of the structure, in contact with the aqueous environment.

The folded form of the protein typically consists of lengths of amino acid that are in a characteristic configuration, such as the α helix or β sheet, forms that we saw in rather sketchy form in chapter 6. But the entire protein does not consist of one long helix or sheet. Rather, sections with one or the other of these forms are connected by turns in the chain, so that the helices and sheets can come together to form the final folded structure. The fumarase reductase structure shown in figure 6.9 (p. 116) provides a good example.

The amino acids along the chain must be spaced in such a way that the hydrophobic groups occur at appropriate intervals. Figure 7.6 depicts a length of protein chain in the α helix configuration, showing only the atoms that make up the backbone of the chain and the side groups attached to the C_α atoms of the chain. Notice that the chain has a natural pitch such that a complete turn occurs about every four amino acids. If the helix section is part of a folded protein, it is likely that one side of the helix faces toward a nonpolar environment, and the other side faces toward the aqueous environment. You can see this in figure 6.9: The helices are positioned so that one side faces inward and the other toward the outside. If the hydrophobic groups, marked in figure 7.6 as NP, are to be mainly oriented toward the same side, they must occur about every three or four amino acids along the chain. There is no precise rule at work here, but

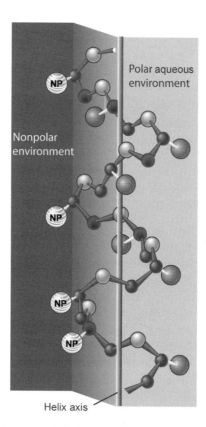

Figure 7.6. A section of α helix chain, showing only the backbone atoms and the side groups attached to the C_α atoms. In this example, the hydrophobic side groups, labeled NP, occur at intervals along the chain such that they line up on the same side of the α helix.

the general idea is that there must be some sort of spacing of the side groups along the helix to provide a predominantly hydrophobic aspect to one side of the helix and a predominantly polar aspect to the other.

An amino acid chain made up willy-nilly from a selection of amino acids is very likely to be nonsensical in terms of the language metaphor discussed earlier. Such a chain will generally not fold, nor will it even be likely to form a helical or sheet structure of significant length. The proteins that occur in nature have evolved over the ages to structures that have the requisite kinds of amino acids in an acceptable order to stabilize the folded form. Nature has already done a lot of work to ensure that the potential for achieving a stable, folded form is there. So if we grant that proteins have been designed through natural selection to adopt a stable, functional, folded form, we are left with one big question: How does the protein find its way from the random, unfolded form to the folded form in which it can carry out the function for which it is designed? The scientist thinks of this question in terms of the location metaphor for a change in state. What is the nature of the pathway connecting the unfolded and folded states?

The Levinthal Paradox

A small conference was held in March 1969 at a conference center of the University of Illinois, "Mössbauer Spectroscopy in Biological Systems." Most of the speakers talked about proteins containing iron because those materials lend themselves to study via Mössbauer spectroscopy. We needn't concern ourselves with what that is. Cyrus Levinthal had something else on his mind. In his talk, titled "How to Fold Graciously," he raised an interesting paradox and challenge.[10] The protein amino acid chain can rotate at each amino acid unit, and typically there are, say, three stable positions of rotation about each bond (recall figure 7.1). In a chain of one hundred or so amino acids, even though the protein can move from one possible rotational configuration to another very rapidly, it would take eons longer than the known age of the universe for a protein to take up all its the possible configurations. Yet we know that proteins fold rapidly, typically in times of a second or less. This means that an unfolded protein cannot possibly find its proper folded configuration by a trial and error exploration of all possible configurations.

At the time of Levinthal's talk it was commonly thought that proteins might find their way from the unfolded state to the folded state in a step-by-step process. Various elements of the folded form would click into place as the protein in its random motions stumbled upon the right combinations of angles. This metaphor is represented rather crudely in figure 7.7, as movement over a surface, commonly called an energy landscape. A similar drawing graced the pages of the *New York Times* on March 25, 1997, in an article on protein folding. The figure was accompanied by this text: "Metaphorically, the process (protein folding) is like a marble rolling through a hilly landscape, randomly trying various paths until it comes to rest at the bottom. In a similar way, a protein tries out many different arrangements until it settles into a stable form."[11]

The metaphor has considerable charm, but as Levinthal noted in 1969, a random motion over the surface won't do the job; folding would simply take too long. Before we look further into how to deal with this problem, let's consider the general character of the metaphor depicted in figure 7.7. It is yet one more example of the location metaphor for change: A change in state is conceptualized as motion from one location to another. To further understand

Figure 7.7. A model of an energy landscape. The "system" (marble or protein) moves over the surface, exploring various valleys and eventually arriving at a low-energy state that may or may not be the lowest point on the surface.

the particular application in figure 7.7, we need to decide what the various directions mean.

The height or depth of the surface is a measure of the energy of the system, which can be a marble, a single protein molecule, or an assembly of protein molecules. Once the marble starts downhill, it acquires kinetic energy, that is, energy of motion. Suppose it rolls into a valley that is so deep that the marble doesn't have enough energy to climb the opposite side of the valley. It then just gets stuck in that valley. The marble has no way of getting an influx of energy to enable it to surmount the valley wall. But protein molecules can acquire energy through continual contacts with their surroundings. If the valley is not too deep, a protein molecule that landed in a valley might thus acquire enough energy to escape. However, if the protein moves into a deep valley, corresponding to a particularly stable form of the protein, even though it is not the correct folded form, it may also get stuck there and remain in a nonfunctional, misfolded state.

But what actual change in the condition of the protein corresponds to motion to the right or left, as opposed to forward and backward? The two horizontal dimensions somehow correspond to changes in the protein molecules. One possibility would be to think of motions from left to right on the surface as changes in the sum of all the Ψ angles (see figure 7.1) and motion in the other horizontal direction as representing the sum of all the ϕ angles. But that won't work because there are many different combinations of individual Ψ and ϕ values that total to the same quantity. Another possibility is that the two dimensions might be measures of "fitness" of the Ψ and ϕ values, with the optimal fitness value corresponding to that for the folded form. For example, one horizontal dimension might correspond to the fraction of ϕ angles that have the values characteristic of the native state. However, this may not be a property of the protein molecules that corresponds to anything measurable or that we can readily compute. At the level of detail implied by the metaphor, we must be content with the general idea that the horizontal dimensions represent "progress toward folding," just as in the simpler chemical example given earlier it represented progress of the reaction. Even without a specific idea of what the horizontal dimensions might mean, the landscape metaphor resonates with our embodied experience of deliberate movement from one location to another to arrive at a particular destination.

Levinthal's paradox, as it came to be called, suggests that the protein can't be just randomly moving around on the surface until it happens to fall into the uniquely correct low-energy spot. The surface is much too extensive, dimpled with many minima of varying depths. The chance of finding the proper folded form by random movements is very small. Furthermore, it is also possible that before finding the correct folded form, the protein might fall into a low-energy depression and be stuck there. The metaphor of a random pathway to folding is just not viable. Levinthal suggested that there must be a small number of favored pathways that the unfolded molecules can follow usually unerringly to arrive at the folded state. Attractive as this suggestion seems at first, it doesn't seem to lead to a workable model for folding. With so many possible configurations open to them, how can the protein molecules get started down one of the few favored pathways? And what is to keep the protein on the favored pathway? In recent years the models for protein folding have continued to use the landscape metaphor, but from a more global (speaking metaphorically) perspective. We will look briefly at one such approach.

Rugged Energy Landscapes and Funnels

Let's see where we can get by giving up the idea that there is a single or only a few viable pathways for the protein to fold. We need to keep in mind that a solution of a protein contains a *lot* of protein molecules, "billions and billions of 'em," as Carl Sagan would say. Each of these molecules in its unfolded state is flopping around, buffeted this way and that by its contacts with the surroundings. The chains twist and turn as rotations around the angles ϕ and Ψ occur (figure 7.1). At times these rotations bring parts of the chain into contact, and an interaction between the atoms of the two different chains occurs. It may happen that the curling up of the chain is such that the contacts between the two parts of the chain are just the ones that lead to a stable form, say the α helix. Those contacts, which characterize the folded, native state, we can call native contacts. In terms of making progress toward folding, those are the good kind. On the other hand, parts of the chain that are separated by many amino acid units along the chain might also come into contact and form some kind of attractive interaction. These "nonnative" contacts are not along a useful pathway toward folding. The protein must undo them before it gets properly folded.

From this point of view, the energy landscape for the proteins is filled with various valleys that correspond to many, many such nonproductive states. José Onuchic and his colleagues call this a "rugged," or "rough," energy landscape: "In the general case, the energy of any compact conformation . . . is a sum of random interactions that give rise to a rough energy landscape like the Alps."[12] Because there are many protein molecules in the solution, all engaged in traversing this rugged energy landscape, many parts of the landscape will be traversed by one molecule or another. What makes the process move in the direction of folding is this: Some forms of native contacts are established rapidly early on, giving rise to sections of the protein that have the native contacts that must be present in the final folded form. For example, sections of the chain may form an α helix. Why? In large measure, because in the course of evolution, sequences of amino acids along the chain conducive to helix formation have been selected. Once a section has formed, it tends to stay in that configuration. Because these sections of helix are stable, the energy of the protein as a whole is lowered as they are formed. This means that the folding landscape is funnel-like. But the shorter sections of helical structure that form must be related to one another in some way to enable the protein to move with some efficiency toward further folding. As more and more local sections fall into place, the protein molecules achieve lower energy, with the final folded, native structure at the bottom of the funnel.

Imagine taking a vertical slice through an energy landscape such as that shown in figure 7.7. One such slice for the rugged energy landscape, with its funnel-like character, is shown in figure 7.8. Notice that the sides of the funnel are not smooth but have valleys in them. The protein molecules work their way down the funnel toward the final state, falling into these little valleys and then escaping from them as they acquire enough energy through interactions with the surroundings. This drawing has been adapted from the review by José Onuchic, Zan Luthey-Schulten, and Peter Wolynes. Of relevance for our analysis of its

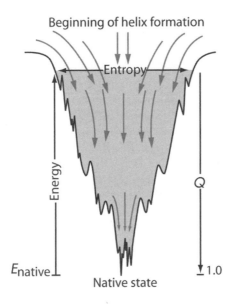

Figure 7.8. A sectional view of a protein folding landscape. The landscape is visualized as "rugged" and having a funnel-like shape.

metaphorical content, greater stability means lower energy, a basic metaphor common to all the landscape scenarios.

Notice that there is a second label, Q, for the vertical scale in figure 7.8. Q is defined as the fraction of native contacts in a structure. When we have a completely unfolded state, Q is zero. In the folded, native state, Q is 1. The protein molecules thus move from an initial value of zero for Q in the completely unfolded form to a final value of 1. In the metaphor represented by figure 7.8, the vertical direction can be taken to represent two different but related quantities. This metaphorical trick works here because Q and energy vary in roughly parallel fashion.

Recall that we have had some difficulty in pinning down a precise meaning for horizontal movement over the energy landscape. What does the horizontal dimension refer to in figure 7.8? Remember that we are looking only at a slice through the surface. To get the full picture, you must imagine a funnel narrowing at the bottom. The protein molecules can approach the center of the funnel from any direction. In the early stages the protein can partially fold rather quickly, through the formation of local sections of native structure. But this means that the structure is getting more ordered. Entropy, as we saw earlier, is related to the idea of order. In the simplest way of thinking about it, entropy decreases as order increases. The horizontal dimension, then, is a measure of the order contained in the protein structure. Remember from the earlier discussion that we assign the most ordered state to be the native, or folded, state. Thus, the bottom of the funnel is taken to be the most ordered state. As protein molecules approach the ordered, folded state, they approach a minimum in entropy. In terms of the horizontal dimensions, the folded state is a kind of "ground zero." The horizontal distance from that reference point is a measure of the disorder or randomness that remains relative to the native state structure. In this way the abstract entity called entropy is metaphorically represented as a linear scale.

The rugged energy landscape model of protein folding, with the concept of a folding funnel, takes the form of a complex theory with extensive mathematical development.[13] Quantities such as the energy and entropy of the system can be computed, in principle at least, in terms of the theory. Figure 7.8 is an attempt to convey its essential elements in the form of a diagram. The elements of the diagram relate to embodied concepts that we all carry with us: movement over terrain, a landscape made up of hills and valleys, movement downward toward a state of lower energy, and so on. The metaphor of a funnel-like surface that can be approached from any direction is an attempt to get around the shortcomings of a model in which only one or a few pathways toward folding are operative. In addition, the proposal of early, fast folding to get to partially folded structures helps to deal with the issue of how long it takes the molecules to attain the folded state.

Summary

In our excursion through the world of protein folding we have encountered several novel metaphorical elements. The idea of protein folding is itself metaphorical, drawn from everyday experiences with the act of folding many kinds of macroscopic objects. The significance of protein properties important for folding, such as amino acid identity and sequence, is understood in terms of a language and information content metaphor. The idea of hydrophobic and hydrophilic types of side groups along the protein chain, and their tendencies to associate with groups of the same kind, is adapted from experience at the macroscale with the immiscibility of oils and water. Most importantly, the folding process itself is seen as movement over an energy surface, a complex metaphor built from the orientational metaphor "More energy is up" and the generalized location metaphor in which changes in state are conceptualized as movement from one location to another.

The various models for protein folding that have been developed over the past several years have provided an impetus to new experimental studies. These have uncovered previously unnoticed aspects of the folding process, such as fast initial steps. Although recently developed models hold promise, however, scientists are still a long way from a fully satisfactory theory of protein folding. Furthermore, it must be admitted that we have been considering a somewhat artificial world. Studies of unassisted protein folding typically are conducted in vitro, that is, in solutions that are much simplified from the conditions that obtain in cells. In the in vitro studies, each protein molecule is unlikely to encounter another protein molecule in its movements in the solution during the folding process. Furthermore, the proteins chosen for study typically are smaller than average, which means as a general rule that they will fold more rapidly.

In the in vivo world of the cell, things are very different. The solution is crowded with many different proteins and a host of other chemicals. In this environment, proteins in the act of folding are likely to encounter another protein molecule during the time it takes them to fold. Such encounters often are fatal for completion of the folding process. Partly folded proteins tend to stick together and eventually form clumps that precipitate, that is, drop out of the solution. Over the course of evolution nature has worked out strategies for dealing with this problem. In the next chapter we'll look at the metaphors we humans use to understand those strategies.

8

CELLULAR-LEVEL METAPHORS

In this chapter we examine aspects of biological entities that are more complex than individual molecules such as proteins. We will see that the metaphors called on in conceptualizing these more complex systems draw heavily from the social domain. The characteristics of social institutions and the interactions between individuals in society that lead to cooperative behavior, competition, and conflict are mapped onto the complex relationships seen at the cellular and molecular levels in biological systems. We begin with a bird's-eye view of the cell and then focus on a few specific processes within it.

The Cell

Robert Hooke, whom we met in chapter 4 as the inventor of the air pump used by Robert Boyle, was a remarkably versatile maker of instruments and a keen observer of nature. In 1665, in *Micrographia,* he reported a wide range of observations, including some made with a primitive microscope of his own devising. When he examined an "exceedingly thin . . . piece of cork," a material obtained from the bark of certain trees, he saw "a great many little boxes."[1] He called the boxes "cells" because they reminded him of the small rooms, or cells, occupied by monks in monasteries. Thus, one of the most fundamental units of biology, the cell, received its name through a metaphorical association with a social institution.

Hooke later looked at living plants and found that the cells of those materials were filled with fluid. Not long afterward, in 1675, Antonie van Leeuwenhoek saw for the first time single-celled organisms, which he called "animalcules," in pond water. These connections with living matter did not immediately lead to an understanding of the fundamental place of the cell in nature. The outlines of the thick cell walls in fresh plant materials became clearly visible as microscopes improved. However, it took a long time to come to the realization that animal matter is also composed of cells. It was not until the 1830s that Theodor Schwann observed for the first time that cartilage contains cells that resemble the cells of plants. From then on it became increasingly evident that individual cells, and the vital processes occurring within them, are the indispensable, fundamental basis of living systems.

We need to have some information and ideas about cells to talk about the metaphors scientists use in the study of biological systems. Figure 8.1 is a simple schematic illustration of some features of eukaryotic cells that are important for our discussion. A great many other components are omitted to keep the discussion as simple as possible. The eukaryotic cell type is characteristic of plants and animals. It differs from the simpler prokaryotic cell, which is characteristic of bacteria, mainly in the fact that it has a nucleus. There is a great deal of variability among eukaryotic cells: Plant cells differ in important ways from animal cells, nerve cells differ from cartilage, and so on. Nevertheless, cells are alike in many important respects.

The nucleus of the cell is defined by the nuclear envelope, a double membrane. The nucleus contains the cell's DNA, stored in the form of chromatin, a large molecular structure formed between DNA and proteins. The nuclear

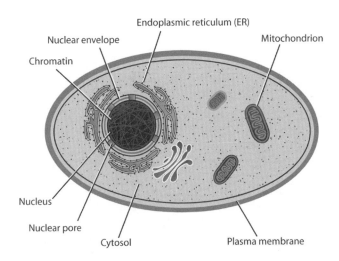

Figure 8.1. A schematic illustration of a eukaryotic cell, showing selected components.

envelope is riddled with nuclear pores, which are elaborate structures made up from proteins and RNA.[2] These structures are somewhat like the "channels" described in chapter 2 but are generally larger and more complex. They have a cylindrical opening down the middle that permits the free exchange of water, ions, and smaller molecules. Passage of larger molecules, such as proteins or RNA, is regulated; that is, certain conditions must be met before those kinds of molecules are passed into or out of the nucleus.

The endoplasmic reticulum (ER) consists of a series of membranous tubes and channels. The ER is the site at which much of the cell's materials are synthesized, most notably the thousands of proteins that facilitate a multitude of processes within the cell. In the mitochondrion food molecules are converted into high-energy chemical substances that the cell uses to drive a host of energy-requiring processes.

The cytosol is the fluid outside the nucleus. It contains myriad dissolved substances and is the medium in which all the distinct bodies of the cell, such as the nucleus and mitochondrion, are suspended. Everything outside the nucleus—the cytosol and everything in it—is called the cytoplasm. Not shown in figure 8.1 is the cytoskeleton, a network of protein fibers that, among other things, gives the cell its shape and determine cell movement.

Keep in mind that there is no typical cell. A nerve cell in your lower back sends a long fiber down your leg all the way to your foot, a distance of perhaps a meter. A cell in the pancreas, on the other hand, is round and on the order of 3×10^{-3} cm (about a thousandth of an inch) across. The two types of cells are built to carry out very different functions, and their shapes and chemical compositions are correspondingly very different. Still, there are commonalities in the workings of all eukaryotic cells.

The cell is the site of a great deal of activity. Scientists have identified hosts of substances and molecular-level structures, have discovered the kinds of chemical processes occurring in the various cellular compartments, and have learned about the transport of molecules and molecular assemblies within the cell. For example, experimental work might reveal that a particular cellular component, say a protein, is produced at a certain stage in the life of the cell and in a particular place. At a later time that protein might become active, facilitate a particular cellular process, and then become inactive. Constructing a model of the cell through synthesis of many such studies is an important goal in cell biology.

Would it be sufficient to simply catalog all the cellular processes? Suppose we could write down all the chemical reactions going on in the cell and somehow contrive a display that would show which are connected to which. Would that give us an understanding of the cell? Perhaps at one level such a model would be explanatory of the cell as a whole, but it could not be readily conveyed from one person to another, nor would it capture a sense of how the cell functions as a single, complex entity. Many properties of the cell, such as its characteristic life cycle, emerge as distinct properties in their own right from the manifold concurrent and closely related individual processes.

Understanding the whole in terms of the multitude of parts can be achieved only if we have a way of conceptualizing the whole. The many individual sets of information can make sense only in terms of a mental model of the cell as an organized entity. For such models the scientist draws on experience with multifarious entities in the macroscopic world. The cell has inputs and outputs, in the form of substances that diffuse through the cell's outer wall. It has budgets of various kinds; for example, the amount of matter entering and leaving must be kept in balance. Energy inputs in the form of food substances that provide energy sources must balance the cellular processes that consume energy. Metaphorical representations that reflect these cellular properties are drawn from the social domain, in which many agents with differing characteristics and goals go about their business, sometimes in competition, sometimes cooperatively, creating in the process a stable entity.

The cell is often seen metaphorically as a factory. We think of a factory as a building or complex of buildings in which things are manufactured. Raw materials go into the factory, finished goods come out. Factories need an energy source; the mitochondrion is seen as the energy plant. Within the factory the employees carry out specialized functions, coordinated to ensure the efficient operation of the whole. To see how extensively this metaphor is used in biology and to appreciate some of its entailments, let's consider a sample of text from a recent issue of *Science,* in which a special section was titled "Frontiers in Cell Biology: Quality Control." The introductory page of this collection of papers begins as follows: "Cells are the basic building blocks of living organisms, and the cell can be pictured as a very complicated factory of life. In order to maintain an effective internal regime and to prevent inappropriate attack by external factors, the cell needs quality control mechanisms to identify, correct and prevent mistakes in its ongoing processes."[3]

The term "quality control" is widely used to describe measures taken to ensure that the products of a multistep process, typically a factory operation, are as error-free as possible. In the field of cell biology, it was first applied in 1989, to describe the process of sorting newly synthesized proteins in the ER. (Recall that the ER is the site of synthesis of most of the cell's proteins.) It turns out that misfolded and incompletely assembled proteins are frequent products of protein synthesis in the ER. The cell needs a mechanism for recognizing and separating out the defective products from the good stuff. We will discuss shortly some aspects of how this sorting occurs. Notice that even the idea of "sorting," of distinguishing and separating the error-free from the defective, hinges on having a metaphorical framework in which to think about experimental observations that do not in themselves convey such ideas. For example, the laboratory scientist might observe that certain proteins present in the cell form associations with other proteins and in so doing influence their behavior and cellular fates. Observations of this kind are interpreted in terms of teleologically rich metaphors in which the cellular components are engaged in purposive actions. To further illustrate the pervasiveness of such metaphors, we turn to a look at these mediating proteins.

The Chaperone Concept

We mentioned earlier that DNA, the genetic material of the cell, is present in the nucleus of the eukaryotic cell in the form of a large assembly called chromatin. Chromatin in turn is composed of chromosomes, which consist of a repeating chain of subunits called nucleosomes. The nucleosomes are made up of a group of protein molecules, called histones, that bind to a section of DNA. (Yes, this is a little bit like Dr. Seuss's *The Cat in the Hat.* To help clarify matters, figure 8.2 diagrams the assembly of chromatin.)

In 1978 R. A. Laskey and coworkers reported a study of the first step in figure 8.2, assembly of the nucleosomes. They found that when they attempted to reconstitute nucleosomes from separated DNA and histones in a solution outside the cell,[4] they did not obtain nucleosomes; instead, a precipitate (a solid) formed. This meant that reconstitution of the nucleosomes had failed. On the other hand, when they added small amounts of a homogeneous mixture made from the cell type they were studying, there was no precipitate on mixing, and nucleosome formation occurred. These experiments are diagrammed in figure 8.3. Laskey and coworkers concluded that some factor in the cell homogenate led to the ordered interaction of histones with DNA to form nucleosomes.

The cell homogenate contains nearly everything that is present in the cell. What among all the multitudinous cellular components could be responsible for preventing precipitation and somehow facilitating the formation of nucleosomes? A succession of experiments led the researchers to conclude that the active agent is a protein. When extracts of the cell homogenate that were highly enriched in the active protein were used in place of the cell homogenate (figure 8.3), precipitation was prevented and nucleosome formation occurred. They were able to ascertain that the active agent binds to histones. Here are the authors' conclusions:

> Therefore we propose that histones are organized into nucleosome precursor complexes by an acidic factor which has a "catalytic" role in the assembly process. The protein we have purified clearly binds histones and transfers them to DNA to form nucleosomes. . . . We suggest that the role of the protein we have purified

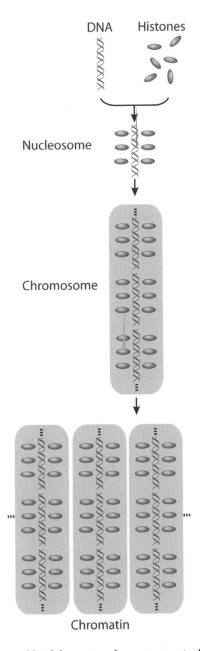

DNA   Histones

Nucleosome

Chromosome

Chromatin

Figure 8.2. Assembly of chromatin, a large structure in the cell that contains the genetic material of the cell, beginning with histones and sections of DNA.

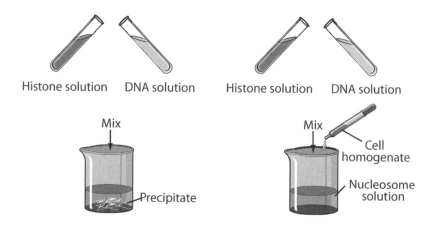

Figure 8.3. Illustration of the experiments demonstrating the presence of an agent in the cell extract that promotes formation of the nucleosomes from histones and sections of DNA. In the absence of the cell extract (left), no nucleosome formation occurs.

is that of a "molecular chaperone" which prevents incorrect ionic interactions between histones and DNA. Thus by binding histones and neutralizing their positive charges the nucleosome assembly protein could prevent nonspecific ionic interactions and allow only selected interactions to occur.[5]

The details of this quote need not concern us. The key points are the use of the term *molecular chaperone* in conjunction with the verbs *bind, transfer, neutralize, prevent,* and *allow.*

The *Oxford English Dictionary* lists several definitions for the word *chaperon* (which in keeping with contemporary practice we will spell as *chaperone*). Historically it referred to a hood or cap, worn by either men or women. In later use, it referred to a person, especially an older woman, who serves as guide and protector for a young, unmarried woman in public. The term may have earned this usage via the metaphorical meaning that the experienced older woman shields her charge from undesirable contacts much as a hood shelters the face. Laskey's use of the term derives from his sense that the active protein that promotes assembly of nucleosomes from histones and DNA does so by preventing inappropriate interactions between charged parts of the molecules involved. Superficially, his use of the term *chaperone* is a nice example of appropriating a term from the social domain to name a newly discovered function of matter at the molecular level. But more is involved than simple catachresis, or naming. Laskey and coauthors are depending on the scientists who read their paper to draw on their understanding of the chaperone social function to grasp the role of the newly discovered nucleosome assembly protein.

The chaperone metaphor can do real work. By calling on an understanding of the social function of a chaperone, it suggests a function for the newly discovered nucleosome assembly protein and even points toward how that function might be carried out. But there is more; like any good evocative metaphor, it has the potential for use in interpreting other observations. Nearly a decade after Laskey's article, in 1987, John Ellis suggested that the term *molecular chaperone* be applied more generally to describe a class of proteins whose function is to ensure that the folding of certain other proteins and the assembly of other proteins into larger, more complex structures occur correctly.[6] The idea is that molecular chaperones prevent incorrect folding and incorrect assembly by binding to the proteins on which they are acting. Recall from chapter 7 that when proteins fold, the hydrophobic, or "water-hating," parts of the chain are tucked into the interior, and the hydrophilic, or "water-loving," parts face out into the solution. Unfolded or misfolded proteins are left with exposed hydrophobic sections. The chaperone protein could act by recognizing these sections and binding to them. In the bound form the proteins are prevented from aggregating with one another, and their correct folding is facilitated. Binding is only temporary; the final folded protein, or final multiprotein assembly, does not contain the molecular chaperone, which, having done its job, moves off. Ellis's move enlarged the scope of the chaperone concept to include a potentially large number of proteins with a range of functions in the cell.

The chaperone metaphor began as a mapping of an idea from the social domain onto the perceived function of a molecular component in the cell. The application by Laskey to the nucleosome assembly factor was followed by the extension suggested by Ellis to proteins that play a role in facilitating the folding of proteins. A very large number and variety of proteins have now been identified as chaperones. It seems likely that most proteins in cellular environments need a chaperone function to attain their native states.[7] By looking at a few examples of application of the molecular chaperone metaphor, we can see how a productive metaphor takes on a life of its own. The range of examples close to the original application expands. In addition, the metaphor has continued to enlarge as new functions in cellular biology are characterized as chaperone-like.

Heat Shock Proteins

One of the earliest elaborations of the chaperone concept involved the so-called heat shock proteins. You will remember that proteins are normally in their lowest energy form when folded, as described in chapter 7. But the folded state is not much more stable than an unfolded or partially folded state. For this reason, many proteins are quite sensitive to temperature. Environmental stresses such as excessive heat cause some folded proteins to become denatured, that is, to lose their biological activity by leaving their native state. The unfolded proteins then may interact with other unfolded protein molecules to irreversibly form aggregates, as when an egg is boiled. A class of proteins called heat shock proteins operate by binding to proteins that come at least partially unfolded during heat or other form of stress. In so doing, they prevent aggregation and facilitate return of the protein to its correct folded form. It appears that some heat shock proteins are able also to promote the disassembly of proteins that have been damaged by stress, causing them to unfold and then refold correctly.

The heat shock proteins are ubiquitous in both prokaryotic and eukaryotic cells but are particularly important in plants, where temperature variations may be extreme. Think of the heat stress that must occur in cells in the leaves of corn plants growing under the Midwestern summer sun. Notice that to function properly, the heat shock proteins must be quite robust themselves and immune to denaturation. In many cases they appear only when the cell is subjected to stress. The earliest studies of heat shock proteins were carried out before the chaperone concept became widely accepted. Only later was it seen that heat shock proteins are a type of chaperone. A newly established metaphor not only stimulates research that gives rise to an elaboration of the original idea but also can initiate a reinterpretation of earlier work.

The Role of Chaperones in Quality Control

Quality control occurs at several stages in the cell's life and in various parts of the cell. Chaperones are a major element in certain quality control processes. As mentioned earlier, chaperones are active in the ER (figure 8.1), where protein synthesis occurs. It turns out that protein synthesis is subject to a fairly high error rate. As much as 20 percent of the proteins synthesized are defective in some way. If some mechanism were not in place to deal with these defective proteins, the cell would soon stop working. Chaperones assist in folding newly synthesized proteins and assembling them into larger units. By binding to incompletely folded proteins, the chaperones hold up the transport of defective molecules and thus prevent their movement out of the ER into other parts of the cell until they are in their native states.

Proteins that are damaged in some way, such as those that are misfolded because of a mutation in the gene that gave rise to the protein, may aggregate with other misfolded or otherwise defective proteins. Normally, aggregation is bad news for the cell. Indeed, some diseases are thought to result from undesirable protein aggregation. To prevent aggregation, defective proteins may be "rescued" by chaperones or destroyed by other proteins that degrade them to the basic amino acid building blocks. Wickner and coworkers call these two processes "protein triage." The language the researchers use in this instance is quite revealing of the underlying metaphor:

> The dictionary defines triage as "sorting and allocation of treatment to patients." The patients in this case are cellular proteins. The first level of triage must be identification of the proteins that are damaged and require treatment. . . . Once damaged proteins have been identified, a second level of decision must be made: Can the patient be saved? Chaperones or chaperone components . . . should have the first opportunity to correct misfolded proteins. Hopeless cases in which structural damage cannot be repaired need to be degraded.[8]

Clearly, the underlying metaphor here is that of a medical facility. This is but one more instance of the point that multiple metaphors often conceptualize distinct parts or functions of a given entity. The idea of quality control, which arises from the overarching metaphor "The cell is a factory," gives way to a more specialized metaphor, "Protein processing is triage." Importantly, both are mappings from complex social domains onto cellular and molecular processes. In each case, interpretations of the observational data are colored by viewing the data in light of the metaphor used.

Prions: Proteins as Infectious Agents

In 1997 the Nobel prize in physiology or medicine was awarded to Stanley B. Prusiner for his discovery of prions and his efforts to prove that prions represent a new biological principle of infection.[9] The award was considered controversial; some biological scientists even today regard his ideas as largely unproven. Prusiner assigned the name *prion* to proteinaceous particles that he believed were responsible for causing scrapie, a brain disease found in sheep and later studied in hamsters. Scrapie is related to a group of other diseases that cause a wasting of brain tissue. This family of diseases occurs in humans as the rare Creutzfeldt-Jakob disease. Most recently, a huge outbreak of a related disease, bovine spongiform encephalopathy (BSE), or so-called mad cow disease, in Great Britain has necessitated the slaughter of about a million head of cattle in efforts to eradicate the disease.

Prusiner's hypothesis is that there is a cellular protein, called PrP (for "prion protein"), that exists in healthy cells in a normal form that causes no trouble. However, under special circumstances the normal form of PrP can convert to an abnormal disease-causing form. It is as though PrP has two basic conformational forms, one normal and the other disease-causing. The controversial part of Prusiner's overall hypothesis is that the abnormal form, let's call it PrP°, is capable of acting on other normal PrP molecules, causing their conversion to the PrP° form. The various prion diseases are caused by the subject's acquiring the disease-causing form, which then goes about infecting the organism by causing conversion of normal PrP to PrP°. Prusiner's hypothesis amounts to a claim for an entirely novel means of infection, one that does not involve replication of an invading organism or virus and thus one that does not involve DNA or RNA. Not everyone accepts his theory of the infectious character of prions, but his work and that of others leave no doubt that prions exist and are involved in some way in brain-wasting diseases.

As we have seen, Prusiner suggests that a normal PrP protein is converted into the disease-causing PrP° form by the action of a PrP° molecule, which somehow causes the normal form to convert. There are no known analogs of such a transformation. Prusiner postulates the existence of an unidentified substance, which he calls protein X, which could bind to PrP and facilitate its conversion to PrP° under the influence of PrP°.[10] Protein X also could be involved in cases of Creutzfeldt-Jacob disease that seem to arise sporadically. But protein X is none other than a chaperone, a protein that binds to another protein and in so doing influences its subsequent fate. The hypothesis that there exists a protein X is an extension of the chaperone metaphor to yet another area of cell biology. The metaphor serves here as the basis of a functional hypothesis to account for unexplained observations. It points toward characteristics that should be looked for in the search for the missing protein X. This example reveals how a successful metaphor can serve as more than simply a name for established observations. Chaperones are thought to be involved in other diseases related to protein misfolding, such as cystic fibrosis and Alzheimer's disease.

Further Extension of the Chaperone Concept

Successful metaphors gain currency by successive applications to continually expanding target domains. The chaperone metaphor is particularly versatile because it evokes the idea of a process or function, one that can be seen to extend to a variety of situations. Whereas the earliest uses dealt with proteins that operated by influencing the behavior of other proteins, a recent application of the metaphor has been to a class of proteins, called metallochaperones, that transport certain metallic ions, such as copper, within cells. These special proteins deliver the metal ions to enzymes that need them for their action and at the same time prevent toxic contact of the metal ions with other cellular processes.[11] The metallochaperones differ substantially from the chaperones that mediate protein folding. Rather than binding to other proteins and mediating their folding behavior, they are involved in transport and delivery of a nonproteinaceous material. However, viewing their role in part as one of "safe delivery," by preventing contact of the metal ions with the general contents of the cell, evokes the idea of a chaperone function. Use of the chaperone metaphor thus has the effect of emphasizing this latter function at the expense of others.

Sides of a Coin: Reductionism and Teleology in Biology

Watson and Crick's discovery of the structure of DNA ushered in the final phase of one of the major transitions in the history of science. Biology has metamorphosed from a largely inductive science concerned with the behavior of whole organisms and biosystems to one that emphasizes the functional meaning of processes occurring at the cellular and molecular level. Even those who study the behavior of bees in nature or the distributions of shrews in northern forests routinely use molecular biological tools in their work. This shift is called reductionism, the attempt to account for the properties of living systems in terms of physical and chemical principles. It is roughly analogous to an attempt to understand the properties of an automobile by learning the workings of all the parts of which the automobile is constructed. Of course the meaning of an automobile is not found in knowing just the properties of all the parts taken individually, without regard to their functioning in the automobile as a whole. In the same way, we can't say that we convey the meaning of a brick wall merely by giving a detailed description of bricks and mortar. The organization of all the parts of the assembled automobile into a working car and the organization of the bricks and mortar into the components of a finished wall impart functionality that is missing from the components. The organized whole possesses emergent properties, those that go beyond the properties of the components that make up the whole.[12] The more complex the system, the more these emergent properties define the system, making it what we recognize it to be.

As we noted earlier, the cell is very complex. It contains all the genetic material of the organism of which it is a part; it may have 40,000 or 50,000 different protein substances carrying out various chemical processes: Molecules are being transported about, individual chemical reactions are being turned off and on, and so on. The scientist can hope to achieve an understanding of every chemical process that occurs in a cell, at a fundamental chemical and physical level. That is indeed one goal of the reductionist program in cellular biology. The molecular level of understanding, in terms of chemical reactions, energy surfaces, and physical processes such as diffusion, is grounded in metaphors and gestalts that are products of the scientist's most basic experiences with the physical world. But explanation at the molecular level alone does not convey an understanding of how the cell works or even of how myriad subdivisions of the cell work. To conceptualize complex systems composed of multiple parts that interact with one another, with interdependences based on transports of matter, chemical conversions, and energy consumption and production, the scientist must draw on experiences in the macroscopic world that have those characteristics. Those experiences occur in the social domains of life.

Some philosophers have worried a great deal about teleological explanations. A logical empiricist approach does not allow explanations that impute purpose to events that occur in nature, most especially to events occurring in the microscopic domain of molecules, atoms, and cells. According to this view of science, the only legitimate statements are those that provide a true account of reality, that is, that correspond exactly to things as they are in nature. Theories related to cell biology therefore would have to consist simply in recitations of the observed chemical and physical processes occurring in the cell, the order in which they occur, and so on. Such a theory would have no predictive power other than by induction; that is, one might be able to infer what an undiscovered system might be like from the behavior of known examples.

Biological scientists evidently do not go about their work with the constraints of logical empiricism in mind. In a time in which biology has become relentlessly reductionist from an experimental perspective, metaphor has become a common tool in understanding observational data and communicating about them. We saw in chapter 3 some examples of teleological explanation. One does not need to understand the underlying biology to appreciate the thoroughly teleological underpinnings of biological explanation in these further examples from the recent literature:

> Now a study . . . suggests that protein aggregates can directly damage cells by hijacking a cellular quality control mechanism.[13]

> In response to apoptotic stimuli, TR3 is translocated from the nucleus to the cytoplasm, where it targets mitochondria to induce cytochrome c release.[14]

> The NMDA-type excitatory glutamate receptor regulates these activity-dependent processes in part by controlling the entry of Ca^{2+} into neurons, which then activates signaling pathways that orchestrate neuronal development.[15]

Of course, no one literally ascribes purpose and human-like intent to molecules and cells. Teleological ascriptions of the kind exemplified here are simply metaphorical ways of conceptualizing complex systems in terms that correspond to human experiences. The important point is that such metaphors evidently are a necessary part of the scientist's understanding of the world under study.

Biology today reveals more forcefully than any other area of science the essential role of metaphor in scientific reasoning and communication.

The realization that metaphorical reasoning is ubiquitous and essential brings to the fore many interesting questions:

> Can we formulate a philosophy of science that recognizes the contingent, embodied, and experiential basis of scientific reasoning and understanding of the world?

> What are the origins of scientific creativity? Armed with the recognition that most of scientific reasoning is embodied and grounded in metaphor, can we do specific things to stimulate creativity?

> What limits to scientific reasoning and communication follow from recognition of the central role of metaphor?

> How is communication between scientists affected by cultural differences that present individuals with different social experiences?

We will attempt to address at least some of these questions in chapter 10.

9

GLOBAL WARMING

In previous chapters we have seen many examples of the ways in which scientists use metaphor in their attempts to understand the world. Simple, fundamental concepts, such as time, quantity, and energy, are understood in terms of directly emergent, embodied experiences (e.g., time is a linear dimension extending backward and forward from the present; energy is a surface). Particular simple physical entities and systems, such as the electron, atoms, and molecules, are conceptualized in terms of experience with macroscopic objects encountered in the everyday world (the electron is a particle, or a wave; atoms are billiard balls). To conceptualize more complex systems that have interrelated parts, scientists draw on metaphors from the domain of social experiences (the cell is a factory; chaperone proteins assist other proteins in folding). In summary, the source domains for the metaphorical representations that scientists use are determined by both the nature of the observational data and the perceived complexity and interrelations of the parts of the system.

In this chapter we examine the roles that metaphor plays in understanding and communicating about changes in Earth's climate. This topic encompasses a host of questions, each of which is complex in its own right. Understanding of climate and climate change involves analysis and integration of ideas from a variety of sources and judgments about how the parts fit together to form a coherent whole. Changes in Earth's climate could have enormous consequences. Many scientists who study these matters foresee human impacts on the global climate over the course of the twenty-first century that will profoundly affect all of humankind. Should we all be as concerned as they are? How reliable are their predictions? What, if anything, can we humans do to control climate change?

One of the big questions regarding climate change is whether some part of the change occurring now and projected for the immediate future is anthropogenic, that is, caused by humans. It turns out that this is not an easy question to answer. The conclusions people want to draw about climate from various observational data are not generally free from uncertainties. Because the implications of the observations and their interpretations are so momentous, the results of scientific work are widely followed with great interest. Significant climate change might adversely affect the global economy, food supplies, health, and many other aspects of life. For these and other reasons, human institutions and people throughout the world have a stake in what science has to say.

We will see that even the simplest questions we ask about Earth's climate have complicated answers. Ideas that seem at first glance to be plainly literal actually turn out to be metaphorical in nature. Our concern in this chapter is not so much with an analysis of the metaphorical content of individual models used in trying to understand climate change, although we will do some of that. Of greater interest are questions such as these:

How are large, overriding metaphors shaped from component metaphorical parts?

Apart from the meaning and significance of terms such as "global temperature," "climate change," "global warming," and "greenhouse gas," what uses are made of the ideas these terms conjure up in advancing competing points of view? How do arguments generated in the context of scientific work get translated into arguments in the larger political and social world to persuade, to support established interests, or to color public opinion?

Let's begin by examining some of the basic science underlying questions of climate and climate change.

Climate

Climate is generally defined as the prevailing or average weather conditions of a place as determined by temperature and other changes throughout the year and averaged over a period of years. Notice that the term *climate* is generally used in reference to a place or region. Thus, the state of New York has a different climate than the Sudan. If the term *climate* were applied to an area as large as the United States, only the most general things could be said, such as that it is a temperate climate as opposed to tropical or subarctic. When we talk about "global climate change," we are not intimating that the entire planet has a common climate but that a set of related climatic changes is occurring over most or all of the planet. A global climate change might produce a wetter climate in one locale and a drier climate in another, but the two changes are traceable to a common origin.

There is abundant evidence that Earth's climate has undergone many changes over time. The historical record, as compiled from observations of sediments, shows that much of the earth's land mass in the northern hemisphere has been repeatedly covered with glaciers. The glaciers appear to have come and gone with periods of about 100,000 years. The most recent major glacial episode reached a peak glacial time, called the Last Glacial Maximum, about 18,000 years ago, and then ended rather abruptly, by about 10,000 years ago. In terms of geologic time, the current interglacial period, called the Holocene, has been a mere blink of the eye. However, it is during this very blink of the eye that humankind has prospered, multiplied, and rather thickly colonized much of the available land mass.

The climatic conditions during past periods can be deduced from a variety of techniques, including examination of fossil records and tree rings from very old trees, analysis of ice cores from old ice, such as the Antarctic ice cap or the Greenland ice cap, and sediments from ocean beds. The particulars of the techniques scientists use to deduce temperatures from study of such samples need not concern us. However, it is invariably the case that a model of some kind, a theory that relates the observations to the presumed conditions at the time the samples are laid down, is needed to deduce temperature from the observations. Thus, estimates of temperature in past times are fundamentally grounded in metaphor. This does not mean that they are unreliable; often a particular model can be cross-checked with other means of estimating temperatures and good agreement obtained. In other cases, laboratory-based experiments provide some assurance that the assumptions of the models are sound.

What about the more immediate geologic past? There is evidence that the climate in many places has changed significantly over the past several thousand years. Many of the changes appear to have been global in character; that is, related changes have occurred at several places on the planet during the same time period. But climate change in even a fairly large region might not reflect global changes. For example, during the period C.E. 1000 to 1300, called the Medieval Warm Period, the climate in significant parts of the northern hemisphere appears to have been warmer than usual. In this case, however, evidence such as data from tree rings suggests that a corresponding warm period was not experienced in the southern hemisphere.[1] Most climatologists today think that the Medieval Warm Period, as reflected in the climate of Greenland and northern Europe, was not part of a global climate change.

Although the climate in specific regions varies from year to year, the evidence from several sources indicates that the global climate is growing warmer. Jerram Brown, a biologist who studies Mexican jays in the mountains of Arizona, has found that the date on which the females lay their first clutch of eggs has grown earlier over the past twenty years or so. In analogous studies, twenty migratory bird species in England have been observed to expand their ranges north over the same period and to lay their eggs earlier in the spring. Field biologists in Europe have noted that the range of nonmigratory butterflies has been extending northward in recent years by up to 150 miles.[2] Glaciers are retreating in almost all regions of the world.[3] The arctic ice is diminishing; measurements made from nuclear submarines over the past half-century show that the ice at many specific locations has been growing steadily thinner.[4] Furthermore, the extent of the arctic ice has diminished.[5] Finally, recent analyses of oceanic temperature data indicate that the world's oceans have warmed by about 0.1°F in the past half-century.[6] This may not sound like much of a warming, but it represents an enormous amount of stored heat, transferred from the atmosphere to the oceans.

These several observations, and many others, collectively constitute evidence that Earth is growing warmer. But evidence based on these kinds of observations, though providing a qualitative sense of global warming, does not tell us how much atmospheric warming is occurring and how it compares with past warming episodes. For a more comprehensive measure of warming we need temperature data. But how does one measure the temperature of a planet?

The Concept of Global Temperature

When we talk about Earth's temperature as it relates to climate, we are referring to the temperature of the atmosphere at or near the surface. But the surface temperature varies greatly at any given moment over the planet. What meaning can "global temperature" have? When we measure human body temperature at just a single place (e.g., under the tongue), we take it for granted that this measured temperature is representative of the body as a whole. Although body temperature may vary a little with location, the value at a particular place, such as under the tongue, serves sufficiently well as an indicator of average temperature. By contrast, because the temperature at any given time varies substantially over the surface of the earth, we can't expect that a single measurement anywhere will serve as an indicator of the planet's temperature.

Of course, it is possible to make up a definition of something we can call the global temperature. For example, we could say that the temperatures at a large number of specified places all over the planet, each averaged over a twenty-four-hour period, could be averaged in some way to produce a single value that we define as the global temperature for that period. The problem is, there are many equally plausible ways to define a global temperature in this manner. No single one of them corresponds to a unique literal reality we can call the global temperature. Any potential candidate definition of the global temperature would have to be judged in terms of how well it met our sense of what a global temperature should be. The idea of a single quantity that is in some sense the temperature of the planet is necessarily metaphorical; it does not correspond to any literal observation.

Granted that the idea of a global temperature is metaphorical, what defined quantity should be used to represent it? In practice, the temperatures at locations on land are measured at frequent intervals and averaged to give a daily value. These daily values are averaged over a month to produce an average monthly temperature at a particular site. The surface of the planet is divided into areas, say 5° in latitude and 5° or 10° in longitude. The monthly temperature averages from each station in a given area are averaged together to give an average value for that area. The values for all areas are then weighted according to the size of each area and averaged to give a grand average for the entire planet. Some areas are well represented by temperature data, others are too remote or inaccessible to provide comparable levels of quality and quantity. In addition, we must keep in mind that nearly 71 percent of the planet's surface is covered with water. The temperature values for oceanic areas are based largely on ship reports, which tend to be sporadic, and are seldom measured for more than a few days within any given area. Does the process seem a little complicated, a little less precise than one would wish? It is! To further complicate matters, the quality and number of temperature measurements has been changing over time. Even after taking this into account, the global temperature estimated for past years may not correspond well to present-day estimates. Nevertheless, this estimated temperature of the planet is the quantity watched with extreme interest from year to year for signs of variations in its value.

To reprise, there are two points of interest here. One is that global temperature, a quantity used to represent the temperature of the planet, is a metaphor. There is no literal temperature of the planet. The idea of global temperature is not made literal simply by making up a definition of the term. Regardless of how we define global temperature, the thing that it is commonly thought to represent, the surface temperature of the earth, is metaphorical. It follows that the concept of global warming, which derives from the concept of global temperature, is metaphorical as well. Second, we have to keep in mind that this metaphorical quantity is not determinable with great precision because of measurement limitations. There will be more on this point later when we come to the use of satellite data to estimate global temperature.

Data of the sort we have been discussing have been available only for the past century or so and only in a few areas even for that period. What about temperature estimates for earlier times? A variety of instrumental methods provide proxies for temperatures in earlier times. For example, it is possible to use data from tree rings to estimate temperature. The data from tree rings that overlap with the modern era can be used to correlate the tree ring data with temperature measurements based on thermometry. Then, with the tree ring data calibrated in this way, the method can be applied to measurements on tree rings that extend back in time as long as perhaps 2,000 years. Similarly, various methods have been used to study ice cores from old ice, such as from the Greenland and Antarctic ice caps. The annual layering of the ice for many thousands of years into the past can be readily detected. The analyses carried out on these ice cores to yield temperature data are based on a theory of how the measured quantity varies with temperature. The measurements themselves are subject to uncertainties and potential for error. In addition, the theory connecting the measured results to temperature may be oversimplified or wrong in some respect, thus adding to the uncertainty. Comparisons with temperature measurements based on other kinds of data, such as analyses of seabed sediments, are important to ascertain whether the same temperature variations are evident in more than one set of data. Uncertainties associated with such comparisons may include the precise determination of the time period. Finally, because the measurements are taken at a few specific places on the planet, the question arises whether they are representative of the planet as a whole.

The Origins of Climate Change

We know from the historical record that the planet has gone through many great changes in climate. The presence of fossil remains tells us that areas that are now deserts must at an earlier time have harbored water, plants, and animals. The presence of rich oil deposits in what are now frozen polar regions tells us that at one time these areas must have supported abundant plant life. The geologic records point to climate changes that are huge in comparison with the climate changes of recent times. It is sobering to consider that during the most recent glacial period, extending from about 50,000 years to about 10,000 years ago, much of the North American continent was covered with ice to great depths. The ice retreated from the area around Chicago only about 12,000 years ago. Milutin Milankovitch hypothesized in 1920 that the longer-term changes that trigger glacial epochs are related to variations in the earth's orbital motions about the sun. Periods of 22,000 years, 41,000 years, and about 100,000 years, computed from observational data on the earth's orbit, can be associated with these variations. However, we are interested in climate change occurring over much shorter times. In thinking about the factors that might have driven climate change over the past thousand years or so and that might operate during the next century, we can forget orbital variations. This leaves three other potential sources of change: volcanism and aerosols, solar radiation, and atmospheric greenhouse gases.

Volcanism and Aerosols

When Krakatoa, a volcano in Java, erupted in 1883, it spewed an enormous amount of very finely divided ash particles high into the atmosphere. A few months later, the airborne ash from that eruption had moved around the globe and was detectable in Europe. The tiny particles suspended in air formed what is called an aerosol. The unusually high concentration of tiny particles capable of scattering sunlight was responsible for spectacular sunsets. At that time the only solar observatory in existence was located at Montpellier, France. The record of continuous solar measurements before, during, and after this time shows that the solar radiation reaching the observatory averaged nearly 10 percent below normal for nearly three years. These and other similar observations support the theory that aerosols in the lower atmosphere reduce the flux of solar energy onto the planet. They do this by reflecting solar energy back into space.

There seems to be little doubt that human activity is contributing to increased atmospheric dustiness. The increase in atmospheric turbidity can be seen even in the middle of the Pacific Ocean, far from any generating sources. Measurements during the past several decades at the Mauna Loa mountain observatory in Hawaii have been showing a steady loss in atmospheric transparency. The sources of aerosols are many and varied; they include sea spray, burning associated with clearing of forests, agricultural activities that raise dust, industrial processes such as metal production, and burning of fossil fuels, mainly oil and coal. The latter source in particular has been especially important. When coal that contains sulfur is burned, most of the sulfur-containing effluent that is not scrubbed from the exhaust ends up as sulfate aerosol particles.

There is general agreement that aerosols are most likely to result in a lowering of the earth's temperature because they reflect a portion of the incoming solar radiation back into space. However, depending on the altitude at which the aerosol is dispersed and whether it absorbs solar radiation, an aerosol might also contribute to warming of the atmosphere. To further complicate analysis, aerosols don't cover the earth uniformly; they tend to be concentrated downwind of aerosol sources, such as a volcano, coal-fired electrical power plant, or urban center. Clearly, no experiment can be devised to measure the net effect of aerosols on the atmospheric temperature. The only recourse is to develop a mathematical model that can be employed to carry out calculations to predict the effect of a given level of aerosol presence.

Solar Radiation

The sun is a remarkably stable and constant source of our planet's light and energy, at least over timescales of hundreds or thousands of years. At the same time, it is a complex spherical ball of nuclear reactions, with its own dynamics and variability. Galileo was among the first to observe areas of immense and violent turbulence on the sun. Later observers noted that these areas, called sunspots, recur with a period of about eleven years. Sunspot activity is accompanied by the arrival on Earth of especially high levels of radiation, most of it of the high-energy kind that produces disturbances in broadcasting and wireless communication. Variations in the amount and kind of solar radiation could affect the global temperature. The data indicate that variations in solar irradiance (in effect, the sun's brightness) from year to year are not large.[7]

Greenhouse Gases

Earth's atmosphere serves as a kind of blanket for the planet, helping to confine heat energy at the surface. One of the earliest scientists to recognize that the atmosphere might have such a function was John Tyndall (1820–93), a British contemporary of Michael Faraday. The idea had also occurred to the brilliant French mathematician Jean-Baptiste-Joseph Fourier. Both scientists were aware that the wavelengths associated with radiation of energy from a cool object, such as the earth, are in the infrared part of the spectrum. These are the kinds of heat rays you can feel by putting your cheek close to a warm biscuit or cup of coffee. The two major constituents of the atmosphere, nitrogen and oxygen, have no significant capacity to absorb infrared radiation. However, several naturally occurring minor atmospheric gases, notably water and carbon dioxide (CO_2), do absorb radiant energy in the infrared region. Nearly all water vapor is found close to the earth's surface; at higher elevations where the atmosphere is colder, it condenses out as water droplets or ice crystals. Carbon dioxide is pretty uniformly distributed throughout the atmosphere. It is present to the extent of about 375 parts per million (ppm) of air.[8] Of the two gases, water vapor absorbs the greater amount of infrared radiation. Nevertheless, CO_2 plays a significant role and is the focus of attention these days.

The role of infrared-absorbing gases in affecting the temperature of the atmosphere is most commonly conveyed using a metaphor that owes its origins to a suggestion of Jean-Baptiste Fourier in 1824. Fourier imagined the atmosphere behaving like the glass cover of a plant starter box. The modern version of the metaphor is illustrated in figure 9.1. The glass walls of a greenhouse permit passage of a substantial fraction of the incoming solar radiation, such as visible and some ultraviolet and infrared radiation. The incoming solar radiation is absorbed by the materials inside, which in turn reradiate heat energy. However, the radiant heat energy emitted from inside is in the infrared region of the spectrum. Whereas glass is transparent to much of the incoming solar radiation, it is not very transparent to the infrared. The result is that much of the infrared energy is reradiated back into the greenhouse by the glass. After dark, when the sun no longer is irradiating the greenhouse, the materials inside continue as before to radiate infrared energy. The glass walls of the greenhouse reradiate much of that infrared energy back, so the greenhouse retains some of its heat energy, as compared with the loss of such energy that would ensue if the greenhouse walls were not there.[9] The right-hand side of figure 9.1 illustrates the analogous role played by infrared-absorbing "greenhouse" gases in the atmosphere. Water vapor, CO_2, and a few other minor gases present in the air absorb some of the infrared heat energy emitted from Earth's surface.

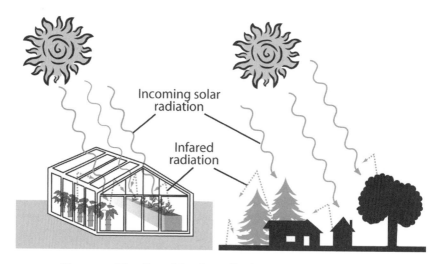

Figure 9.1. The effect of the glass walls of a greenhouse in retaining infrared heat radiation is illustrated on the left. The analogous effect of atmospheric "greenhouse gases," mainly water vapor and carbon dioxide, is shown on the right. These gases, though transparent to visible light, absorb and reradiate infrared radiation, thus keeping a larger fraction of the radiated heat energy in the lower atmosphere.

Some of the absorbed energy is reradiated back toward Earth, thus keeping the lower atmosphere and Earth's surface warmer than it would be otherwise.

The superficial comparison of the action of the atmospheric gases with the glass walls of a greenhouse is inviting, but as is true of metaphors in general, it masks many complex factors that are important in determining whether and just how water vapor and CO_2 produce a temperature effect. Our concern in this chapter is not to dwell on the metaphorical nature of the model itself, however, but rather to explore quantitative models based on it and the ways in which the greenhouse metaphor and related ones are used in advancing differing points of view on social issues.

Because carbon dioxide is a greenhouse gas, there has been speculation that variations in carbon dioxide levels of the atmosphere might have been the cause of past climate changes. As noted earlier, Tyndall and Fourier first made these suggestions in the early nineteenth century. Near the end of the nineteenth century, the Swedish chemist Svante Arrhenius raised the prospect that the burning of fossil fuels (coal, oil, and natural gas) contributes to increased concentration of atmospheric CO_2 and thus might affect the global climate. There are indications from experimental data that in past geologic times the carbon dioxide content of the atmosphere has varied as periods of glaciation have come and gone. Lower levels of carbon dioxide are associated with periods of extensive glaciation.[10] Several lines of evidence indicate that carbon dioxide levels have varied over time. For example, a study of air bubbles trapped in ice cores in Antarctica indicates that the carbon dioxide content of the atmosphere during the period from 60,000 years ago to 20,000 years ago varied from as low as 190 ppm to 220 ppm.[11] Ice core data on carbon dioxide levels during the nineteenth century indicate that the preindustrial level of CO_2, up to about 1850, was steady at about 280–90 ppm. From that time onward there has been a steady increase in CO_2 levels, as shown in figure 9.2. Since 1958 accurate measurements of CO_2 levels, initiated by Charles D. Keeling, have been made at the Mauna Loa in Hawaii, a remote mountain location largely free of transient variations. They show an accelerating increase in CO_2 levels, to the present value of about 375 ppm (figure 9.2). By correlating these accurate measurements with the less precise ice core data, scientists conclude that the present level of CO_2 is easily the highest it has been in the past 420,000 years.

When organic matter decays or burns, carbon dioxide is released. During the past century the burning of fossil fuels, such as wood, coal, gas, and oil, has resulted in exceptional releases of CO_2 into the atmosphere. Over a long period of time, say a few centuries, most of this added CO_2 will find its way into the oceans or be incorporated into new plant growth. However, because these anthropogenic sources are so large and occurring over such a short period of time,

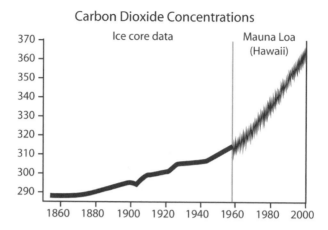

Figure 9.2. Variation in the concentration of atmospheric carbon dioxide over time. The more recent measurements are part of a continuing observation at Mauna Loa Observatory in Hawaii.

the CO_2 is not being absorbed by the oceans as rapidly as it is being added. At present, human society derives most of its energy from fossil fuel combustion. Assuming that this continues to be the case, the atmospheric CO_2 concentration is expected to continue increasing, perhaps at an even faster rate, until about 2050. By then it is predicted to be about 550 ppm, about twice the pre–industrial age value.

Based on the model of CO_2 as a greenhouse gas, it is reasonable to expect that increases in its atmospheric concentration will affect the temperature at the earth's surface. The buildup of CO_2 therefore will result in warming of Earth's climate. Should we be concerned? Will the increase in global temperature caused by the increase in CO_2 level be large enough to produce noticeable climate change? If it is, will that change be truly global in nature or confined to particular regions? Will any changes that ensue be for good or ill? All these questions are of potentially great importance to human society. By burning the planet's store of fossil fuel over a short period of time, we are in effect conducting an unprecedented experiment on a global scale. If there is a chance that the predicted doubling of the CO_2 concentration will produce large climatic changes, should we not take steps now to limit burning of the vast stores of fossil fuel that remain? No laboratory-based experiments can provide answers to these questions. Any attempt to respond will require the development of a sophisticated model that yields quantitative predictions.

Modeling the Climate

Climatologists are challenged to develop a model for the climate of the planet that incorporates the three factors that might affect climate: aerosols, solar irradiance, and greenhouse gases. The model must be an effective simulation of the atmosphere and ocean. It must incorporate what is known of the behavior of the atmosphere and oceans (e.g., circulation patterns and currents) and include interactions of the atmosphere with the oceans and with the surface of the earth. It should be able to predict what climate changes would result if the sun's irradiance changed by a certain amount, a volcanic eruption added a certain amount of aerosol materials to the atmosphere, or the atmospheric CO_2 concentration changed by a certain amount. Models of this kind, called general circulation models, are computer-based mathematical formulations requiring enormous computer capacity. Even with the largest and fastest computers available, the models must sacrifice a great deal of detail to simulate the behavior of the system in any reasonable period of computation. They are so complex and have so many interrelated parts that they do not lend themselves to simple, intuitive reasoning.

Not all scientists who work on these problems have the same view of what the ideal general circulation model looks like. Several such models are in use, based at least in part on independently derived components. They produce various kinds of results that can be compared with observational data. For example, a calculation might be run to simulate the conditions thought to apply to some past climatic regime for which data are available from fossil sources, ice cores, tree rings, and so on. The output of the calculations can be compared with the observational records to determine whether the model can replicate features of past climates and climate changes. If the model holds up well under such tests, one can have some confidence in its ability to forecast future climate changes. Furthermore, the various models can be compared with one another to assess consistency.[12]

There is evidence that the sun's irradiance has varied slightly over the past millennium, enough to account for perhaps 0.2°C (0.4°F) during the period.[13] Variations in aerosol levels do affect the global climate; a higher level of aerosols probably would have an overall cooling effect. On the other hand, the models indicate that it would take higher increases in aerosol levels than are now anticipated to produce significant long-term cooling. This leaves the greenhouse gases as the major variables in the climate prediction model. As already noted, the increase in CO_2 concentration probably will be the major contributor to change in the greenhouse gases. Water vapor levels are important, and they will change. However, water is continuously entering and leaving the atmosphere. The amount of moisture in the atmosphere is determined to a significant extent by the temperature at and near the surface. Considered globally, the contribution from water vapor changes only when the global atmospheric temperature changes. A higher temperature could result from an increase in concentration of one of the greenhouse gases. In this way, a CO_2-induced warming of the atmosphere probably will lead to an increased concentration of water vapor, thereby enhancing the greenhouse warming produced by the added CO_2 alone.

The various global circulation models used to estimate the effect of increased concentrations of CO_2 on climate are in agreement: Increases in CO_2 concentrations will result in global warming. However, the models differ in their assessments of how much warming will occur and where it will be most evident. In 1995, at the request of Congress, a panel of climate modeling scientists developed a statement on the credibility of modeled projections of climate change. They produced a ranked list of the predictions in order of their degree of certainty. They concluded that it is very probable that the surface temperature of the earth will continue to rise through at least the mid-twenty-first century.[14] The best estimate, in the view of the panel members, is that the global mean surface temperature will increase by about 0.5°C to 2°C (roughly 1°F to 3.5°F) over the period from 1990 to 2050. In 2000 a new report was issued by the United Nations–sponsored Intergovernmental Panel on Climate Change, updating its 1995 report. The estimates of global temperature increases from anthropogenic influences remain about the same. However, there is substantial uncertainty in these and other estimates because of unknown contributions from, among other things, aerosols and cloud coverage and the uncertainties about human consumption of fossil fuels.

The obvious question that follows from these results is whether there is any experimental evidence indicating that the global temperature is increasing and, if it is, whether the increase can be tied to the increasing concentration of CO_2. Because climate is sensitive to so many variables, it is difficult to make a strong argument that the increase in CO_2 concentration since 1860 has been sufficient to bring about a clearly discernible global temperature increase. Several studies that have appeared quite recently point to a sharp increase in global temperature during the past few decades. One of these publications describes construction of a global temperature profile for the past millennium, based on a variety of proxy data such as tree ring analyses and ice core samples based on data from the northern hemisphere.[15] The resulting graph of the temperature variation is shown in figure 9.3. The shaded areas about the curve represent the estimated uncertainties of the data. Despite the large uncertainties, a definite

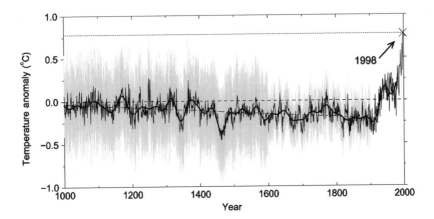

Figure 9.3. A reconstruction of global mean annualized temperature over the past 1,000 years. (Reprinted from M. E. Mann, R. S. Bradley, and M. K. Hughes, "Northern Hemisphere Temperatures during the Past Millennium: Inferences, Uncertainties, and Limitations," *Geophysical Research Letters* 26:6 [1999]: 759, fig. 3A. © 1999 American Geophysical Union)

but slight trend toward lower global temperature seems to have been operating over the past 1,000 years, until the latter half of the twentieth century. At that point there is a sharp increase in global temperature through the year 1998. The authors conclude, "Even the warmest intervals in our reconstruction pale, however, in comparison with *modern* (mid-to-late 20th century) temperatures. . . . Both the past year (1998) and past decade (1989–98) are well documented as the warmest in the 20th century instrumental record. . . . The past decade and past year are likely the warmest for the Northern Hemisphere *this millennium*."

Using a different approach, based on a coupling of modeling with the instrumental climate record, T. J. Crowley has also constructed an account of global temperature variation over the past 1,000 years. He also identifies an anomalously large late-twentieth-century warming and concludes that the bulk of it is consistent with increased concentrations of greenhouse gases. Finally, a recent comparison of observations with simulations using a coupled ocean-atmosphere general circulation model points toward both natural and anthropogenic factors in twentieth-century temperature changes. In agreement with the conclusions of other studies, warming of the past two decades or so can be accounted for only in terms of anthropogenic factors.[16]

These and other studies have of late convinced increasing numbers of atmospheric scientists that increases in CO_2 and other greenhouse gases are producing global warming today and will contribute to still greater warming as more and more fossil fuels are burned. But this conclusion by itself fails to answer the most important questions: To what extent and in what ways will Earth's climate be changed by any warming that does occur? The global climate models make several general predictions with a fairly high degree of reliability. For example, they predict that the amount of sea ice in the northern hemisphere will be diminished, that precipitation will increase, on average, over the globe, and that the sea level will rise. As an indication of the extent of the uncertainty in these predictions, the rise in sea level is predicted to range from 2 to 16 inches by 2050. When it comes to predicting specific changes in weather patterns over regions of the globe, however, the models are not as consistent with one another.[17] About all that can be said with confidence is that changes in the climates of regional-scale areas such as large urban complexes, states, or smaller countries are likely to differ from one another. Climate change will not be uniform over the planet. As a particular example, if sufficient fresh water is added to the upper layers of the North Atlantic Ocean (e.g., by melting of the arctic ice sheet or increased precipitation at high latitudes), the Gulf Stream current that carries warm waters to the western coast of Europe might be disrupted. In that case, western Europe would become much colder, while other regions of the globe grow warmer.

So what can we conclude from all of this? The metaphor of climate change caused by variations in the atmospheric concentrations of CO_2 and other so-called greenhouse gases has been with us for a long time. The qualitative notion that such gases exert a warming on the atmosphere near the earth's surface has been supported by model calculations that have grown increasingly sophisticated over the past half-century. The story remains the same, although the quantitative details vary from one model to another. These quantitative details are not unimportant. If the predicted change in global temperature that results from a doubling of the CO_2 concentration is at the low end of the range of predictions, we may not have to worry much about inimical climate changes. On the other hand, if the temperature change is at the high end of the range, the human race probably is in serious trouble.

In much scientific work, when more than one possible outcome is foreseen by competing models, the models are put to the test of comparison with experimental results. That sometimes can take a long time because the experiments may not be feasible with the tools available or because, as in the case of climate change, a long time is needed for the necessary results to accumulate. Climatologists are in the difficult position of looking for a temperature increase caused by added CO_2 while many other factors might also effect a temperature change.

To further complicate matters, the global temperature, itself a metaphorical idea, is not readily estimated with precision. Data of high quality, collected from uniformly distributed locations on the planet, are not abundant even now and have certainly not been available historically. Only in the past few years have the data (and data analyses) begun to look sufficiently solid to convince many atmospheric scientists that Earth's surface temperature is undergoing an anomalous increase. But not everyone is convinced. Furthermore, the implications of the observed global warming for weather are not widely agreed on at all; many are dubious about the entire scenario. Let's look at the contrarian perspective.

Contrarian Views

Climatologists have known for a long time that the weather is controlled by large-scale exchanges of air between the tropics and the poles. These large-scale phenomena give rise to the counterclockwise movements of air that may lead to hurricanes and typhoons. Many atmospheric patterns are not stable for long periods. Thus, as we learn from the weather reports, the jet stream, a west-to-east movement of air in the temperate zone, moves about and causes frequent weather changes. On the other hand, some large-scale atmospheric and ocean-atmospheric patterns have natural periods of years. Local or regional weather thus depends on atmospheric fluctuations of only a few days' duration, on seasonal effects, and on multiyear patterns. For example, a pattern of wetter weather in the southeastern United States has been ascribed to the El Niño, in which the eastern South Pacific Ocean is warmer than normal. But it might be caused by a secular (in one direction) shift caused by a change in some fundamental global climate factor, such as CO_2 concentration.

The fact that global circulation models cannot make very precise predictions leaves plenty of room for alternative explanations of observed weather trends. Furthermore, many of the observational data themselves, on which the assertion of a global temperature change is based, are also open to challenge. One group of critics questions whether in fact there has been a change in global temperature during the past half-century. Although such a change is indicated by the surface measurements already alluded to, some say the temperature estimates based on satellite data tell a different story. The satellite instruments, which view the atmosphere from above, detect the temperature of a section of the atmosphere extending from the surface to an elevation of about 6 miles. Over the past couple of decades the atmosphere sampled by the satellite measurements shows little or no warming, a finding at variance with that implied by the surface temperature measurements. Climate modeling skeptics point out that the models do not predict a difference in warming between the surface and the lower atmosphere. The fact that a difference apparently exists casts doubt on predictions based on global climate models.

The first question that must be addressed is whether the apparent difference in temperature trends is real. Recall that surface temperature estimates are problematic in terms of completeness and accuracy. The satellite measurements have their own potential errors. For example, slight changes in the orbits of the satellites over time were improperly corrected for until recently, causing systematic errors in the temperatures derived from the measurements. This example illustrates a point so often overlooked in discussions of how science works. The establishment of a "fact" requires data that are sufficiently precise and free from systematic error to serve the purposes of the investigation. Second, the observational data must correspond unambiguously to the quantity being measured. The "fact" of global warming during the past few decades is not yet established on either count. There is a serious conceptual problem in connecting the metaphorical idea of a single number called global temperature to either surface or satellite measurements. There is no literal, one-to-one correspondence between any set of measurements and the concept that we hold of a global temperature. When we try to map two distinct sets of observational data onto a single metaphorical entity, there is bound to be a problem. The attempts originate at least in part in the mistaken idea that global temperature is a literal, objective reality to which some set of measurements corresponds.

There is a broad range of views among atmospheric scientists on whether the predictions of global climate models are reliable. It is possible for scientists to entertain such a wide range of views because the models are not readily testable against relevant experimental data. Even the most avid supporters of global climate models admit that quantitative predictions for times well into the future are highly uncertain. In a recently published report dealing with the uncertainty of the models, the authors have this to say:

> Forecasts of climate change are inevitably uncertain. . . . Here we assess the range of warming rates over the coming 50 years that are consistent with the observed near-surface temperature record as well as with the overall patterns of response predicted by several general circulation models. We expect global mean temperatures in the decade 2036–2046 to be 1–2.5 K (1.8–4.5°F) warmer than in pre-industrial times under a "business as usual" emission scenario. . . . Unlike 50-year warming rates, the final equilibrium warming after the atmospheric temperature stabilizes remains very uncertain, despite the evidence provided by the emerging signal.[18]

In short, it is possible to make an estimate of global warming over the next fifty years, but the range of values of this quantity is pretty large. The increasing uncertainty at still longer times arises because no one is sure about the rate at which humans will still be generating CO_2 or how rapidly it will be leaving the atmosphere by one pathway or another. Forecasts of changes in regional climates based on global climate models are even more open to question than the quantitative predictions of temperature changes. There is no general agreement as to what the global climate changes mean for each region of the globe.

Metaphors and Public Policy

Global climate modeling and prediction have been characterized over the past few decades by disagreements and outright disputes. This is not an unusual phenomenon in science, particularly when predictions are based on complex models that are sensitive to the particular assumptions that go into their formulations. The difficulties in obtaining reliable, accurate data on past and current climate only add to the uncertainties. As we have seen, global climate models are metaphorical in nature. They are constituted from a host of individual models for the various factors contributing to climate. There is a model for how the atmosphere exchanges with the oceans, for how water vapor is distributed, for transport of the atmosphere from the tropics to higher latitude, and so on. Each model is incorporated into the larger model, with terms that couple it to other parts in what is thought to be the appropriate way. Because of the interdependence of the various components of the overall model, the eventual outcomes in the form of surface temperature profiles or patterns of water vapor distribution are sensitive to the assumptions embodied in each of the contributing models. The fact that the predictions cannot immediately be compared with observational data puts them on quite another footing from a theory that can be tested by going into the laboratory and carrying out a series of experiments under carefully controlled conditions.

If the differences in approaches to predicting global climate were confined to a segment of the scholarly research community, the controversies would arouse no unusual level of interest. But these studies are about something that could affect everyone in a future world society. What humans are doing today has the potential to change the climate far into the future. The practices that may so dramatically change climate are strongly entrenched and will not readily yield to change. As a sense of urgency has grown over the past few decades, nations have begun to explore policies that would lead to reduced production of greenhouse gases, particularly CO_2.

In the United States, many environmental groups have placed global warming high on their agenda of urgent issues. In response to pressures from various interest groups, Congress passed the Global Change Research Act in 1990. Among other provisions, it mandates periodic assessments of the effects of climate change on water supplies, human health, agriculture, coastal areas, and marine resources. The United Nations established the Intergovernmental Panel on Climate Change. The assessment reports of that panel have been influential in setting the tone of international discussions of steps that might be taken to limit CO_2 emissions. In 1997 more than 150 nations met in Kyoto, Japan to formulate international policies for limitations on fossil fuel consumption on the grounds that CO_2 emissions are the major driving force for global warming. The sorts of policies needed deal with how limits in CO_2 emissions are to be established, particularly in distinguishing developing nations from highly industrialized ones, or steps a nation might take to offset a portion of its emission total (e.g., planting new forests that would sequester CO_2). The outcomes of those discussions, called the Kyoto protocols, have been under discussion for the past several years. They call for a group of thirty-eight industrialized nations to reduce their CO_2 emissions in the 2008–12 period. In 2000 the nations reconvened at the Hague in the Netherlands to put the finishing touches on an international agreement intended to be signed by the nations of the world.

Even before the Kyoto meeting, there was talk of reducing CO_2 emissions. President George Bush agreed with European leaders at the 1992 summit in Rio on a goal of reducing emissions to the 1990 level by 2000, and President Clinton later reaffirmed it. But goals are one thing, concrete actions are another. Mandated limits on fossil fuel consumption could have enormous impact on the economies of the highly industrialized nations. In the absence of dramatic technological change or vigorous pursuit of energy conservation mechanisms, energy consumption runs roughly in parallel with gross domestic product. The Kyoto protocols as they apply to the United States could require a 5 percent reduction in CO_2 emission by 2010. There has been no legislative action that would move the American economy in the direction of lower fossil fuel consumption other than mandates on overall gasoline mileage ratings for automobiles. Most members of Congress seem to be opposed to the Kyoto protocols because they are deemed to threaten the vitality of the U.S. economy.

Strong forces are arrayed on both sides of the issue of whether anything should be done now to forestall effects, uncertain as to nature and extent, that will not be experienced until well into the future. Actions taken on an effective scale will necessarily have major social and economic consequences. Those who think we should take action now advocate, among other things, a reduction in fossil fuel emissions, commitments to reforestation, and development of technologies for sequestering CO_2 (e.g., by pumping it underground into old oil fields). Others believe we should do nothing. The advocates for action include most environmental groups, some government agencies, and many researchers in the scientific fields associated with the issues. Those who advocate only watchful waiting include free market libertarians, various industry groups, and scientists who are skeptical of the models, which they do not believe are supported by the experimental data.[19]

The Transformation to Social Metaphor

It is not my aim here to dissect all the positions held by various individuals and groups or to analyze their reasons for adhering to the positions they maintain. Indeed, by the time this book is in your hands, much will have changed. Whatever changes in understanding may occur, however, it will remain true that the metaphors that underlie the science are transformed as they find their way into the larger public discourse. As with any useful metaphor, global warming maps onto aspects of the observational world in ways that make for a coherent story. To be sure, there is potential for disagreement about the precise nature of the mapping onto the observational domain, that is, about which data best correspond to the concept. In any case, however, global temperature and global warming are central elements in the models that incorporate them. When these metaphors are used in public discourse, however, their character changes. In this larger arena the term *global warming* does not bring to mind the underlying scientific ideas. Instead, it has become the name for something that resonates because of its commonly held, or folk, beliefs.

Global warming is an especially interesting case of such a shift in character, for two reasons:

> The future climate is predicted using complicated models of the global climate that yield uncertain results. The model builders believe that they have included all the important terms, but it is difficult to be sure. Beyond this, even if all the relevant factors are included, the necessary simplifications in terms of realistic inclusions of land-air and ocean-air interactions, the role of clouds, the coarseness of the grid over which the calculation must be carried out, and other such factors all lead to uncertainties. Finally, even when the calculations predict an increase in global temperature, we are still uncertain about region-by-region climate change. In short, there is room for a lot of disagreement over whether science can reliably predict future climate.

> The potential for global warming caused by human influences is a scientific issue that draws great interest in the larger society because of the potential economic and social ramifications.

Global warming exemplifies a class of scientific problems with immense social portent. It is complex and difficult, yet we need reliable predictive analysis.[20] Models for dealing with it, such as global circulation models, have much in common with those encountered in the social sciences, where models are also used in attempts to simulate complex aspects of the world. In economics, for example, one might construct a model to predict the prices of particular economic units, say corporate bonds. It would necessarily be formulated in terms of a great many components, each largely metaphorical in nature. The model would yield outputs that map onto things in the real world (e.g., corporate bond prices, supply of bonds, interest rates, inflation). In effect, it would say, "Given the following scenario with respect to the elements of the model, this is what will happen to corporate bond prices in such and such a future time period." Such models have some of the characteristics of the gedankenexperiments we talked about in chapter 1. One sets up a hypothetical universe with a particular set of properties, then visualizes what will happen if a particular set of circumstances arises with respect to the ingredients of the model. Global circulation models predict what will happen if a particular scenario develops in terms of CO_2 concentration or the concentration of a certain type of aerosol. The predictions of such models are uncertain because the models are incomplete in ways that we can't avoid and because the scenarios we input are uncertain. Just as bond prices may be affected by behaviors of bond buyers and sellers in response to global conflicts and a host of other factors not included in the model, global temperature may be affected by climate variables not properly included in the model or by uncertainties in the actual rates at which CO_2 is generated.

In dealing with problems characterized by complexity and uncertainties, the physical and natural sciences are not fundamentally different in character from the social sciences in their approaches to model building and interpretation. The models in each case are formed from metaphors that reflect embodied and social experiences, as appropriate. In the social sciences, however, setting the problem in an appropriate metaphorical context often is a more difficult challenge.[21]

In his book *If You're So Smart: The Narrative of Economic Expertise*, Donald N. McCloskey describes the ways in which stories and models complement one another in conveying the content of economics and other sciences. He draws attention to the complementary roles of storytelling and the metaphorical models that cast stories in analytical terms. The attempt to banish figurative language and thought from any field of science, whether it be physics, chemistry, biology, economics, or political science, is impoverishing:

> The modernist experiment in getting along with fewer than all the resources of human reasoning puts one in mind of the Midwestern expression, "a few bricks short of a load." It means cracked, irrational. The modernist program of narrowing down our arguments in the name of rationality was a few bricks short of a load.

To admit now that metaphor and story matter also in human reasoning does not entail becoming less rational and less reasonable, dressing in saffron robes or tuning into "New Dimensions." On the contrary it entails becoming more rational and more reasonable, because it puts more of what persuades serious people under the scrutiny of reason. Modernism was rigorous about a tiny part of reasoning and angrily unreasonable about the rest.[22]

When scientific issues are seen as public policy issues, the language and terms used in discussions of the science take on new roles. They become tools in persuasion and lose much of their connection with the scientific issues. For example, the term *greenhouse effect* is widely used in the news media to talk about the effect of gases such as CO_2. It is likely that for most people it does not bring to mind a model of how the gases actually operate to cause atmospheric warming; it is simply a name for the role of the gases in causing warming. In this new venue it plays a different kind of role in persuasion. Stephen Schneider relates a story that brings this new role into focus:

> While some atmospheric scientists have advocated dropping this well-heeled term, greenhouse effect is both too entrenched and, even if inexact, not that bad an analogy to what the atmosphere does to trap heat near the Earth's surface. Ironically, perhaps, some environmental activists have also advocated dropping the phrase, not because it is a physically inexact analog. Rather, they fear that since a greenhouse is a warm and friendly place for life, the term carries too benign an image with respect to the human enhancement of the atmosphere's heat-trapping capacity. They prefer global heat trap.[23]

I recently received in my mail a solicitation from Physicians for Social Responsibility. The outside of the envelope carried this message:

> Global Warming.
> It's Real.
> It Kills.
> What can you do?

Here are two quotes from the letter inside:

> Scientists now believe that *warming ocean waters could cause the average global sea level to rise by 13 to 20 feet over the next century, submerging the Florida Keys, and all of South Florida and Florida's coastal cities.*

> More and more scientists now agree with what PSR has known all along . . . the greenhouse effect, caused primarily by our increased use of and dependence on fossil fuels, is responsible for the rise in temperatures all around the globe.

For a very different take on global warming, we turn to an op-ed piece in the

Wall Street Journal by Arthur and Zachary Robinson titled "Science Has Spoken: Global Warming Is a Myth."[24] These authors have this to say about the potential consequences of burning fossil fuels: "What mankind is doing is moving hydrocarbons from below ground and turning them into living things. We are living in an increasingly lush environment of plants and animals as a result of the carbon dioxide increase. Our children will enjoy an Earth with twice as much plant and animal life as that with which we are now blessed. This is a wonderful and unexpected gift from the industrial revolution." On one hand, global warming is seen metaphorically as a social evil. It does great damage to society; it kills. On the other hand, the Robinsons see global warming as either a myth or a nonthreatening turn of events. In both cases, the metaphor that began life as an element in climate model building has been transformed. To the extent that it now conveys images associated with political and economic considerations, global warming has become a social metaphor. And this is by no means a unique instance of such a transformation. Metaphors associated with genetically engineered plant and animal products, strategies for combating the AIDS epidemic, and contraception come to mind as further examples.

In chapters 4 through 9 I have described several case studies that illustrate how scientific explanations of observed properties of the natural world are built on metaphorical models. The models are created through a process of reasoning that is tied inextricably to bodily capacities, everyday experiences of physical surroundings, and gestalts drawn from experience in the social domains of life.

The examples were chosen to exhibit a wide range of subject matter, beginning with the simplest and most ancient. The idea of atoms as the indivisible ultimate particles from which the physical world is constructed has been part of Western culture for more than two millennia. To the present day, our concepts of the atom, whether as a single undivided entity or as one with interrelated constituent parts, are based on models that draw on experiences with objects in the macroscopic domain. There is no single concept of the atom that serves all scientists' needs. Today there is talk of seeing atoms with the aid of powerful new experimental tools. But the most sophisticated experimental tools for extending our sensorimotor capacities to the microscopic domain provide data that can be interpreted only through the agency of metaphorical models. This is but another way of saying that in modern science, most observations are theory-laden. For example, a physicist might use a device called a photon counter to measure very low levels of radiant energy. Any observations made with this device are interpreted in terms of the model of radiation as a stream of photons first put forward by Einstein in 1907.

And so it goes with molecular models. We "see"

the microscopic world of molecules, from simple ones such as aspirin to large biological molecular complexes, with the aid of powerful experimental methods such as X ray diffraction. But the data obtained in diffraction experiments consist of tables of angles and relative intensities of X rays scattered by the sample. In themselves the data tell us nothing; only through the agency of models and theories can we convert raw observational data into something that makes sense. The three-dimensional structures that prove so useful are obtained by application of theoretical expressions based on metaphorical models for how X rays are scattered and detected.

Much of what is important in our lives has to do with change. Just as we need models for structure in nature, we also need models for processes. In chapter 7 we examined protein folding, an important example of change at the molecular scale. In the general location metaphor, change is conceptualized as movement from one location to another. The language used in science to talk about processes is suffused with imagery associated with the metaphor of change as movement. A common conceptual framework for understanding processes, of which protein folding serves as an example, is that of movement over a surface, perhaps in steps, perhaps over barriers, from one location (state) to another.

The understanding of observational systems that possess several interacting components demands more than application of metaphors drawn only from directly emergent physical experiences. We saw in chapter 8 that talk about the cell, seen as a complex entity with coordinated processes between many subunits, proceeds in terms of metaphors drawn from the social domain. Human organizations with which we all have recurrent, pervasive experiences have the requisite levels of complexity and interactive character to serve as models. The cell might therefore be imagined as a factory, a city, or a hospital. Within the complex, highly organized cell, certain proteins are observed to interact with other proteins in ways that are conceptualized as chaperone-like. The traditional social function of the chaperone is used metaphorically to advance the idea that the chaperone proteins protect, mediate interactions of one protein with another, and correct imperfections in their charges. There is no escaping these strongly teleological, metaphorical devices for interpreting observations in complex systems characteristic of biology.[1]

All of these examples illustrate the thesis that the conceptual foundations of science are thoroughly metaphorical. Presented with new aspects of the world, we humans understand them in terms of deeply ingrained bodily and social experiences that already form the framework for dealing with life on a day-to-day basis. The fact that metaphor is so inextricably a part of the fabric of science also means that it plays many roles. Scientists use metaphorical reasoning to interpret observational data, creating models to account for new observations and to reinterpret older data. Metaphors, once created and put to use in these ways, serve in communication between scientists and between scientists and the public. Such communication depends on explanatory language that conveys the essential ideas with clarity. A good metaphor can also be persuasive. Selling a new idea and receiving credit for it are important for the scientist's goals of achieving recognition.

Metaphors form the key ideas that go toward defining distinct areas of research. In fact, they are the central elements of what Thomas Kuhn famously spoke of as a paradigm. Without attempting to define a term so much used and abused, one can say that a scientific paradigm includes at least a theoretical, conceptual core grounded in one or more metaphors. Metaphor is the vehicle by which ideas and models from one scientific discipline are transferred to another. For example, the idea of a "code" has found its way from information theory into molecular biology; the idea of a "spin glass," important in the study of disordered systems in physics and metallurgy, has been applied in models for protein folding. Metaphors that conceptualize the mind as a computer are clearly traceable historically to the development of digital computers. Obversely, neural nets and genetic algorithms, both fruitful developments in computer sciences, are grounded in metaphors drawn from the biological sciences.

The community of science is situated in the larger society and cannot exist independently of it. The products of science and technology impinge in important ways on peoples' lives. Scientists may also at times make claims that have important social, political, and economic implications. When this happens, as we saw in chapter 9 (dealing with global warming), scientific models become more than just tools for advancing scientific understanding. Where there is a strong, larger public interest, the metaphors of science take on new roles, and their metaphoric character changes. In becoming icons for positions with public policy implications, they surrender their explanatory character.

Counterviews

Many would disagree with views I have expressed in this book, for a variety of reasons. I want here to address what seem to me the more important counterarguments to my positions.

Some would disagree about what constitutes metaphor itself. Theories of metaphor have shown up in abundance in the past few decades, prompted by the emergence of linguistics and cognitive sciences more generally as important areas of study. The first line of disagreement is the fundamental position that conventional metaphors, as distinct from more obviously figurative language found in poetry and fiction, are not really metaphors at all.[2] I do not find the empirical work supporting these claims very convincing, but I am outside my field of expertise. Dedre Gentner and her colleagues have emphasized the importance of mental models and structure mapping (akin to the knowledge representation schemes considered in chapter 2) in scientific discovery and understanding.[3] Although she takes issue with certain of the claims made by Lakoff and his colleagues, her work is generally consistent with the views I have expressed regarding the importance of metaphorical reasoning in science. In any case, I do not intend to make a detailed defense of the theory of conceptual metaphor here, beyond what I said about it in chapter 3 and, in passing, in subsequent chapters. Lakoff and Johnson are excellent sources of material on this issue.[4] In explicating the occurrences of conventional, everyday metaphors as systematic features of language, the theory of conceptual metaphor renders moot many of the historically disputatious issues of what constitutes metaphor[5] or distinctions between different kinds of metaphors. In my judgment the theory provides a very useful, self-consistent model with explanatory power. I hope that the preceding chapters have shown that a host of common linguistic conventions in science consistently reveal the embodied, metaphorical nature of scientists' understanding of the world.

Aside from the narrower issue of whether so-called conceptual metaphors should really count as metaphorical, there is a larger question of whether we gain anything by deconstructing scientific language in this way. I believe that we do, that we learn important things about how scientists reason and communicate. Our descriptive account shows that conceptual metaphors permeate scientific language, that they are used in many different contexts and are drawn from various domains of human experience. Analysis of the language scientists use helps us appreciate the importance of embodied, directly emergent physical experiences in scientists' individual thought processes. We have seen also that scientists' life experiences in social domains strongly inform what and how they observe and reason.

Those inclined toward a stronger version of scientific realism than that which follows from conceptual metaphor theory are not happy with the kind of intersubjective knowing that we argue for. I do not aspire in this modest volume to settle or even materially contribute to long-standing philosophical debates. Scientific realism comes in many different flavors,[6] but nearly everyone agrees on a couple of basic premises. First, there is a real world out there, and it exists independently of our knowledge of it. Second, the great successes of modern science and technology go a long way toward assuring us that we can have stable, reliable knowledge of nature. Flying in a jet aircraft from one city to another and listening to favorite music from a CD provide rich material for reflection on this point. However, it does not follow that scientific theories are true descriptions of things as they really are, of entities possessing unobservable as well as observable properties. The question that exercises us, seemingly without end, is whether our knowledge of the world is "objective," that is, whether we can claim to approach a true or approximately true account of nature, whether it makes sense for us to aspire to know absolute truths that exist independently of us.

The strong realist position assumes that we humans can aspire to attain a literal, universally true understanding of nature. The physicist Sheldon Glashow holds to such a position: "We assert that there are eternal, objective, ahistorical, socially neutral, external and universal truths, and that the assemblage of these truths is what we call Science and the proof of our assertion lies in the pudding of its success."[7]

I find a great deal in Glashow's essay with which to agree, but the obvious successes of science, including those in Glashow's own field of particle physics, do not provide justification for such a sweeping claim. By contrast, a philosophy grounded in embodied realism makes a case that we know the world only in terms of perceptions, categorizations, and reasoning, both conscious and unconscious, grounded in our bodily capacities and life experiences and inherently limited by them. Embodied realism denies that there is a single, absolutely correct description of the world. This does not mean that anything goes, that knowledge is entirely relative, or that normative measures of what anyone can claim to know can't exist. Directly embodied concepts such as basic-level categorization, spatial relation concepts, and event structure concepts, are products of evolutionary human development over a long time. We all share in these, even though our individual experiences vary in detail. Our neural systems have arrived via evolution at a common architecture and *modus operandi*. We are therefore capable of sharing observations and communicating ideas about them, although differing individual experiences in both the physical and social domains preclude our having exactly the same conceptual views.

One strong measure of a model or theory is its capacity for predicting new observations. It has been argued that models or theories that are not at least approximately true cannot be expected to have predictive power. This seems entirely reasonable, but the real issue lies with what one means by "true." Models and theories are anchored to reality through analogical relationships constrained by observation. That is, they are shaped by our particular capacities for observation and reasoning, and their truth content is judged in terms of correspondence with knowledge gained through those capacities, not by correspondence with a mind-independent reality. Furthermore, because observations are themselves theory-laden to varying degrees, productive models and theories—those that are consistent with the data and that successfully predict new observations—are ineluctably related to a larger body of modeling and theory that provides the immediate context of any particular work in progress. In this way, a network of self-consistent models has been constructed over time. Their cumulative capacity to account for observations results in part from the fact that the observations themselves are theory-laden. This doesn't make them mere will-o'-the-wisps. Clearly, to work as well as they do, successful models in science must relate closely to the world that is out there. But considering how we come by our knowledge of the world and the basis of our reasoning about it, the strong realist claims can't be justified.

On reflection, it is difficult to imagine what a strongly realist science might look like. To find a nonloaded, pure observation language free from the taint of metaphor probably would prove impossible, in practice if not in principle. The famous twentieth-century philosopher Ludwig Wittgenstein in his later work raised the issue of "family resemblances" as a counter to the idea that there is an unambiguous mapping from words to their referents.[8] Empirical research in cognitive sciences has amply demonstrated that categorization is a human activity, serving human needs, that bears no necessary relationship to things as they are in the world.[9]

There is no need here to pursue these issues further. It is sufficient for my purposes to have shown for several different kinds of problems in science, originating in diverse scientific disciplines (and therefore in different scientific cultures) and applying to nature at different scale levels, that conceptual metaphor is a vital ingredient in reasoning and communicating about the world, observed "scientifically."

Metaphor and Science Education

The task of the scientist is to understand some particular part of the physical world, that is, to give an explanatory account consistent with all available observational data and subject to testing through additional experimental work. We have seen in this book many examples showing that in reasoning about the world and interpreting observations, scientists rely largely on metaphorical concepts. Ortony refers to this as "constructivist," in the sense that cognition results from mental construction.[10] From the perspective of conceptual metaphor theory, this mental construction of a representation of the world is grounded in embodied reasoning and in experiential gestalts, elements of structured understanding drawn from everyday life. I have argued throughout this book that there is no special kind of rationality that sets science apart from other realms of thought. At the same time, scientific reasoning is facilitated by special methodological protocols and ways of organizing data that would not necessarily be appropriate to other disciplines, such as the law or literature.

Teaching science, whether to nonscientists or novice scientists, involves many components. These include conveying the scope of a particular area of science, the special nomenclature needed to talk about the subject, and perhaps also the arts of experimental work in the field. Most importantly, it involves imparting conceptual understanding and a sense of intellectual excitement about the subject. The creative use of metaphor is a vital element in that process. Unfortunately, the visitor to many science classrooms, whether fifth-grade general science, high school biology, or college-level chemistry, is unlikely to see much attention being paid to establishing conceptual understanding. Instead, the emphasis is likely to be on learning the names of things, on properly classifying them, and on solving numerical problems by plugging numbers into appropriate equations. Much empirical work in science education has shown that although students can memorize names and formulas and can learn to recognize which equations to use in a given situation, they often lack a correct conceptual understanding of the subject.

The theory of conceptual metaphor helps point the way toward good science teaching practice. The first step must be to focus on description: This is the part of the world on which we will be focusing, here are some observations and data. Second, students are invited to use their reasoning powers: What can we make of these observations and data? Can we relate them to things we already know? What sort of explanatory model can we come up with? What does it mean to say that we have explained the observations? What further observations would help us develop a better and more complete model of the system? How can we test our hypotheses, either in direct experiments or via thought experiments that test the logic of what we have proposed? Can we identify one model that seems to best accommodate all that we know of the system? Is the model consistent with other things we know that relate to it?

The teacher's job is to help students learn how to observe and how to answer questions that arise from the observations. The major emphasis, initially at least, should be on a qualitative understanding, one that draws on the students' capacities for embodied reasoning based on their own life experiences. A sound conceptual understanding provides the basis for exploring the quantitative aspects of the subject.

Every science teacher knows that students' embodied experiences often leave them with incorrect ideas about how things work. For example, many people misunderstand the relationship between force and motion, believing that for an object to continue in motion under any circumstance there must be continued application of force. This is a natural enough concept to hold, given that in our world friction or gravitational force inevitably brings motion to a halt unless continued force is applied. A good part of the art of effective teaching involves identifying such misconceptions and finding ways to overcome them. It is not enough to simply assert the correct way of looking at things. Empirical studies have shown that students cannot easily divest themselves of incorrectly learned concepts. Optimally, they should be led to experiences that demonstrate the concept so that it can be correctly understood. When this is not possible, as it often is not, thought experiments and guided analogical reasoning should be used.

Armed with the recognition that most scientific reasoning is embodied and grounded in metaphor, specific things can be done to stimulate scientific creativity, particularly among young people. The scientific imagination could be fostered early in a child's life by promoting thoughtful observation of nature and by encouraging the formulation of hypotheses about the origins of observed behaviors and properties. Children should be encouraged to formulate explanations and to test them by asking "what if" questions that can be addressed through experiments or by logical analysis. Even at the level of graduate education, an attitude of adventurousness about performing new experiments and advancing hypotheses should be encouraged. Actively encouraging creativity in this way offers an attractive alternative (for the graduate students, at any rate) to performing experiments that seem designed as uncritical collections of yet more data, as confirmations rather than tests of hypotheses and theories.

The Social in Science

Many nonscientists and more than a few scientists have what can justifiably be called an elitist image of science.[11] They look upon science as a highly specialized form of intellectual activity and practice, pretty much a closed system, governed by special rules that don't apply more generally. Even when laypersons feel unhappy about a public policy matter that hinges on technical issues, they may not feel competent to call into question the scientific conclusions on which one side or the other rests its position. Public airings of various scientific controversies, such as nuclear waste disposal, environmental cleanup of the Hudson River, production of genetically engineered foodstuffs, and anthropogenic carbon dioxide generation occasion more confusion than enlightenment about how scientists justify their claims. Often, the layperson's sense of confusion is born of more than mere lack of expertise. There is a mystique surrounding the processes by which scientists arrive at their conclusions, a sense that incomprehensible methods have been used, leading to conclusions so wrapped in complexity as to be virtually unassailable. Scientists often actively encourage an attitude of passive acceptance of scientific findings, sometimes because they believe that science has special claims to rationality.

The impression of objectivity in science is enhanced by the conventions of scientific communication. Scientific papers usually are written in language from which the subjective has been excluded as much as possible. The tone of objectivity that characterizes scientific communication, oral as well as written, glosses over the personal, subjective elements that may have played important roles in choices of experiment or interpretations of results. But although their metaphorical, subjective character is disguised by the rhetorical forms employed, the conclusions to which scientists are driven by their work are not unequivocally objective in nature.

In recent decades a growing number of social scientists have become interested in the social aspects of science. The stronger assertions made by some of these scholars have been badly received among scientists. At various times the "science wars" have been quite heated.[12] The view of the role of metaphor advanced in this book helps put some of the contentious issues in these debates into a useful perspective, one that lies somewhere between the strongly objectivist stance of some scientists on one hand and a strongly social constructivist view on the other. Once we see that scientists' views of the world are based on reasoning grounded in directly emergent physical experiences and in gestalts built from experiences in the social domain, the ways in which they interact with one another to form the social structure of science can be more fully appreciated. The many examples provided in earlier chapters show that scientists' observations, and the models and theories used to make sense of them, are not, as some would claim, mere social constructs devoid of bearing on an underlying reality. To be sure, they are "constructed," in the sense that they are the products of individual physical experience with things in the world. It is also true that scientists are influenced by social experiences with the many complex entities that constitute the economic, social, and political life of any contemporary human community. So it is quite correct to say that culture as well as embodied experience shapes the scientist's understanding of the world and influences choices of subjects for study and approaches to the studies themselves. But science is a game of rules. Observations are subject to the requirements of experimental control and reproducibility; theories must relate in reasonable ways to existing or potential observations.

A view of scientific reasoning as grounded in conceptual metaphors is an effective counter to various degrees of epistemic relativism that call into question the status of scientific knowledge.[13] According to the stronger positions of social constructivism, the social environment largely determines what the scientist chooses to focus on, what is interpreted as reality, and the terms in which models and theories are expressed. But by looking at how conceptual metaphor works in science we see a much different picture. Scientists are indeed constrained by their experiences. The major controlling influences are directly emergent physical experiences. Additionally, many gestalts that form important elements in the scientist's reasoning are the products of generalized social experiences, such as those that give rise to the idea of a factory, say, or a journey. These subjective physical and social experiences are largely shared. Indeed, intersubjective recognition of common basic experiences and ways of interacting with the world are just what make it possible to communicate observations and reasoning about them. But it is one thing to argue that the scientist uses his structured understanding of a factory or the postal system in reasoning about observations in cell biology. It is quite another to assert that social pressures born of competition, ambition, or political outlook exert significant influences on scientific outcomes. J. J. Thomson may have been motivated because he was British to think of electrons as particulate rather than wavelike and may have accordingly been influenced in his choices of experimental design. Having made such choices, however, he was not free to place whatever interpretations he chose on his observations. There is simply no plausible way to understand the development of science as shared, cumulative knowledge by supposing that interpretations of observations shaped predominantly by social forces such as competition, ambition, or political outlook can acquire an enduring status or that socially constructed theories can prevail in the face of requirements of consistency and experimental tests. Recent books by Robin Dunbar, Andrew Pickering, and Jan Golinski provide provocative analyses of the relationships between knowledge production and practice in science.[14]

Social aspects of science are evident also in multidisciplinary research groups, in which scientists with distinctly different experiences in science come together to work on problems that may be too complex to be fruitfully approached from the perspective of a single discipline. Each scientist brings to the common effort a background rich in the metaphors that dominate in his or her specialty. When these metaphorical insights are shared, each person in the group has an opportunity to see things in a different light, and creative new possibilities open. Even casual interactions often pay off. As an example, I learned recently of a collaboration born of a conversation between a physiologist and an electrical engineer. The physiologist was having difficulty in developing an adequate theoretical model for ion transport in ion channels of the kind we spoke of in chapter 2. The engineer, whose research interests are in theories of electronic devices, saw the charge transport in very different terms than the physiologist. He identified an analogy to electron transport under certain conditions in semiconductors and was able to suggest a mathematical approach that greatly simplified the theoretical formulation for ion channels.

Metaphor figures prominently in the social processes of communication between scientists and between the scientific community and larger society. Consider, for example, the very beginning of the research process. Just as in politics one must get elected to be in the game, in science you've got to get funded. A large proportion of all research occurring at what academic scientists think of as the frontiers of their disciplines is funded by grant support that must be obtained by application to external funding agencies, such as the National Science Foundation, the National Institutes of Health, and various private foundations. Proposals to these organizations are most typically processed through peer review, that is, review by other scientists who are presumed to act as impartial experts in evaluating both the author of the proposal and the proposed research plan. In addressing these peer reviewers, research proposals are rhetorical documents, designed to explicate, excite, and persuade.

The organizations responsible for allocating grant resources tend to be conservative. Although a high value is placed on originality, there is also reluctance to fund research that appears to be high risk. Reviewers look for connections with what is deemed to be mainstream and feasible. The result is that fashionable science tends to be funded at the expense of other, possibly more creative, riskier ideas. What is fashionable in turn tends to be defined by a select handful of commanding metaphors. For example, today the metaphors of "nanotechnology," "self-assembly," and "combinatorial methods," applied across the research areas of materials, drug discovery, and biological science, are commanding metaphors. An impressively large fraction of new proposals in these fields use these metaphors in one way or another. The power of metaphor to engage the imagination works in this instance, through the social processes of science, to constrain personal choices of research problems.

Because science is a social enterprise as well as the sum of the work of individual scientists, commanding metaphors play important roles in determining the directions taken in research. The metaphors put forth by eminent scientists, those perceived to be leaders in their fields, are more likely to become widely adopted, all other things being equal, than those advanced by lesser known workers. This is yet another manifestation of the "Matthew effect:" "The Matthew effect consists of the accruing of large increments of peer recognition for particular scientific contributions to scientists of considerable repute and the withholding of such recognition from scientists who have not yet made their mark."[15] Robert Merton and his colleagues have extensively documented this effect, which takes its name from a passage in the first book of the New Testament. But regardless of who initiates them, commanding metaphors generally have modest beginnings. We saw in the example of chaperone proteins how an initially sketchy, evocative metaphor, applied to a single, specific set of observations, was picked up by another scientist and then evolved over time to embrace a diverse set of observations in many different contexts. Productive metaphors are neither inevitable nor arbitrary. Because they come to life in the particular cultural milieu in which the individual scientist finds herself at the time, they are at least partly accidental products of the social context in which she works. At the same time, they are shaped by her embodied and other cumulative life experiences. There is no separating the scientist's acts of knowing from her situation in both her physical and social worlds. John Ziman, whose book *Real Science* provides an engaging, comprehensive, and realistic examination of science as a culture, has this to say:

> Scientific knowledge is not just a disembodied stream of data or the books on a library shelf. It is generated and received, regenerated or revised, communicated and interpreted by human minds. Human mental capabilities are remarkable, but also limited. They are also closely adapted to the cultures in which they operate. Many of the characteristic features of science are shaped by the psychological machinery that scientists employ, individually and collectively, in their study of the world. In other words, cognition is the vital link between the social and epistemic dimensions of science.[16]

Final Thoughts

If, as I have argued in this book, we have reason to believe that metaphorical reasoning in science is ubiquitous and essential, interesting questions arise. For example, can there be a philosophy of science that explicitly recognizes the contingent, embodied, and experiential basis of scientific reasoning and understanding of the world? In their ambitious book *Philosophy in the Flesh*, George Lakoff and Mark Johnson begin with three premises:

> The mind is inherently embodied.
> Thought is mostly unconscious.
> Abstract concepts are largely metaphorical.[17]

They assert that these major findings of cognitive science are inconsistent with much of Western philosophy. Because this philosophy bears directly on science, they have things to say about the status of scientific knowledge in light of their view of the embodied mind. I find their arguments reasonably self-consistent, stimulating, and convincing. In the wake of the long and unproductive hegemony of certain forms of empiricism, any philosophy of science that hopes to attract serious attention should take account of science as it is practiced. Embodied reasoning, experiential gestalts, empirical findings regarding categorization, and other results from cognitive sciences must be accorded important roles in scientific thought and creativity.

We might ask what limits to scientific reasoning and communication follow from recognition of the central role of metaphor. If the scientific imagination has its origins in the scientist's embodied reasoning and dependence on experiential gestalts, it follows that the workings of the physical world can be seen only through the lenses of embodied and social experience. There is no purely abstract thought, somehow divorced from our embodiment. Even so-called pure mathematics, which deals with relationships that seemingly bear no direct connection with bodily experience, is fundamentally grounded in our experiences with objects in the perceived world.[18] What limits on the range of scientific inquiry are implied by such assertions? For one thing, we give up the unproductive metaphor that goes by the name "The theory of everything," the idea that there is an ultimate theory of the universe.[19] This quest for the ultimate in reductionism is predicated on the belief that there exists out there, awaiting our discovery, a theory that accounts completely for all the fundamental particles and forces that make up the universe. Should such a theory even exist, however, it would prove utterly useless in accounting for nearly everything that matters. The complexity that we perceive in the real world, even in so simple a system as an atom that contains several electrons, renders "The theory of everything" irrelevant. The emergent properties of systems—that is, those that arise from the interactions of their various parts—are just the ones that make most things we study interesting. Yet these are just the properties that a reductionist theory fails to capture.[20]

If we abandon Platonic notions of ideal forms existing independently of human thought, waiting to be discovered by minds somehow free from bodily constraints, we are left with proceeding just as scientists apparently do: They make up stories about how the world works based on what they perceive of the world through their senses or instrumental extensions of the senses. They test the truth values of those stories by working out the consequences in terms of models, then designing and performing experiments. The stories, the models, are continuously processed through the filter of human experience with the world. The truths we learn about the world are, as the American philosopher Richard Rorty asserts, not something out there waiting to be discovered but truths made by human agency. They are grounded in representations drawn from our embodied experiences and the gestalts that form our structured understanding of complex entities in the everyday world.

In the respect that scientists understand the world in terms of metaphors grounded in embodied and social experience, they are precisely like everyone else who aspires to be creative.[21] In modeling the sea, the scientist uses metaphors that highlight one set of its properties without touching at all on aspects that captured the attention of Herman Melville or Joseph Conrad. The scientist and the novelist use the same capacities of reasoning and imagination in their attempts to understand different aspects of the same reality. When we recognize that scientific reasoning is based on the same kinds of thought processes used in other arenas of thought, that scientists are constrained in their attempts to read nature by the same embodied and social understandings that everyone uses to get along in life, science is really not so mysterious after all. What is mysterious and wonderful is the power of metaphorical thought to call forth the highest exercise of human intellectual powers.

NOTES

Chapter 1: Scientific Thought and Practice

1. K. R. Popper, *The Logic of Scientific Discovery* (London: Hutchinson, 1959).

2. H. Putnam, "The Corroboration of Theories," in *Scientific Revolutions*, ed. I. Hacking (Oxford, England: Oxford University Press, 1981), chap. 3.

3. J. R. Platt, "Strong Inference," *Science* 146 (1964): 347.

4. Platt wrote, "Whether it is hand-waving or number-waving or equation-waving, a theory is not a theory unless it can be disproved. That is, unless it can be falsified by some possible experimental outcome" (ibid., 350).

5. Michael Polanyi is an important but not widely known figure in twentieth-century philosophy and sociology of science. Born in Budapest in 1891, he earned a Ph.D. in chemistry from the University of Budapest in 1915. He taught in Germany for several years, then moved in 1929 to Victoria University in Manchester, England. Although he had made outstanding contributions to physical chemistry, he was excluded because of his national origin from carrying out research during World War II. He turned this circumstance into an opportunity to begin serious study of the intellectual and social underpinnings of science. In 1948 he exchanged his chair in physical chemistry for one in social sciences at Manchester. His single most important work is *Personal Knowledge: Toward a Post-Critical Philosophy* (Chicago: University of Chicago Press, 1958), but also of interest are Polanyi's *The Tacit Dimension* (London: Routledge and Kegan Paul, 1966) and Polanyi and Harry Prosch, *Meaning* (Chicago: University of Chicago Press, 1975).

6. Polanyi, *Personal Knowledge*.

7. Polanyi, "The Creative Imagination," *Chemical and Engineering News*, 25 Apr. 1966, 88.

8. Polanyi, *The Tacit Dimension*.

9. Polanyi, "The Creative Imagination," 93.

10. D. C. Berry, "The Problem of Implicit Knowledge," *Expert Systems* 4 (1987): 144–51.

11. A. S. Reber, "Implicit Learning and Tacit Knowledge," *Journal of Experimental Psychology* 118 (1989): 219.

12. A. S. Reber, R. Allen, and P. Reber, "Implicit versus Explicit Learning," in *The Nature of Cognition*, ed. R. J. Sternberg (Cambridge, Mass.: MIT Press, 1999), 504.

13. Reber, "Implicit Learning," 233.

14. N. Nersessian, "How Do Scientists Think?: Capturing the Dynamics of Conceptual Change in Science," in *Cognitive Models of Science*, ed. R. N. Giere (Minneapolis: University of Minnesota Press, 1992), 3–44; D. Gooding, "The Procedural Turn; or Why Do Thought Experiments Work?," in *Cognitive Models of Science*, ed. R. N. Giere (Minneapolis: University of Minnesota Press, 1992), 45–76.

15. G. Holton, "Metaphors in Science and Education," in *Metaphors of Education*, ed. W. Taylor (London: Heinemann Educational Books, 1984), 91–113; G. Holton, *Einstein, History, and Other Passions* (Reading, Mass.: Addison-Wesley, 1996).

16. Gooding, "The Procedural Turn," 72; see also D. Gooding, *Experiment and the Making of Meaning* (Dordrecht, The Netherlands: Kluwer Academic Publishers, 1990).

Chapter 2: Introduction to Metaphor

1. John Locke, *An Essay Concerning Human Understanding*, ed. P. H. Nidditch (Oxford, England: Clarendon Press, 1975), 508.

2. *Oxford English Dictionary*, 2d ed., s.v. "metaphor."

3. E. Wilson, "Riverton," in *A Comprehensive Anthology of American Poetry*, ed. C. Aiken (New York: Modern Library, 1944), 422–23.

4. R. Jeffers, "Continent's End," *The Collected Poetry of Robinson Jeffers*, Vol. 1, ed. T. Hunt (Stanford, Calif.: Stanford University Press, 1988), 16.

5. T. S. Eliot, "The Love Song of J. Alfred Prufrock," *The Complete Poems and Plays 1909–1950* (New York: Harcourt Brace Jovanovich, 1952), 3.

6. I. A. Richards, *The Philosophy of Rhetoric* (London: Oxford University Press, 1936); M. Black, *Models and Metaphors* (Ithaca, N.Y.: Cornell University Press, 1962); E. F. Kittay, *Metaphor: Its Cognitive Force and Linguistic Structure* (Oxford, England: Clarendon Press, 1987); D. S. Miall, ed., *Metaphor: Problems and Perspectives* (Sussex, England: Harvester Press, 1982); C. M. Turbayne, *The Myth of Metaphor*, rev. ed. (Columbia: University of South Carolina Press, 1970); G. Lakoff and M. Turner, *More Than Cool Reason: A Field Guide to Poetic Metaphor* (Chicago: University of Chicago Press, 1990); D. Gentner, B. F. Bowdle, P. Wolff, and C. Boronat, "Metaphor Is Like Analogy," in *The Analogical Mind: Theory and Phenomena*, ed. D. Gentner, K. J. Holyoak, and B. Kokinov (Cambridge, Mass.: MIT Press, 2001), 199–253; R. W. Gibbs, *The Poetics of Mind: Figurative Thought, Language, and Understanding* (New York: Cambridge University Press, 1994); R. J. Fogelin, *Figuratively Speaking* (New Haven, Conn.: Yale University Press, 1988).

7. D. G. Bobrow and A. Collins, eds., *Representation and Understanding: Studies in Cognitive Science* (New York: Academic Press, 1975); D. E. Rumelhart and A. Ortony, "The Representation of Knowledge in Memory," in *Schooling and the Acquisition of Knowledge*, ed. R. C. Anderson, R. J. Spiro, and W. E. Montague (Hillsdale, N.J.: Erlbaum, 1977), 99–135.

8. G. Audesirk and T. Audesirk, *Biology: Life on Earth*, 5th ed. (Upper Saddle River, N.J.: Prentice Hall, 1999), 74.

9. Gentner et al., "Metaphor Is Like Analogy," 199–252.

10. R. Harré, review of *Metaphor in the History of Psychology*, ed. D. E. Leary, *British Journal of the Philosophical Society* 47 (1996): 141–45.

11. D. Gentner, "Are Scientific Analogies Metaphors?," in *Metaphor: Problems and Perspectives*, ed. D. Miall (Sussex, England: Harvester Press, 1982), 128; E. R. MacCormac, *A Cognitive Theory of Metaphor* (Cambridge, Mass.: MIT Press, 1985).

12. M. B. Hesse, *Models and Analogies in Science* (Notre Dame, Ind.: University of Notre Dame Press, 1966).

13. A. I. Miller, *Insights of Genius: Imagery and Creativity in Science and Art* (Cambridge, Mass.: MIT Press, 2000), 219.

14. Hesse, *Models and Analogies in Science*, 157.

15. M. Bradie, "The Metaphorical Character of Science," *Philosophia Naturalis* 21 (1984): 229–43; M. Bradie, "Explanation as Metaphorical Redescription," *Metaphor and Symbol* 13 (1998): 125–39.

16. J. Ziman, *Real Science: What It Is and What It Means* (Cambridge, England: Cambridge University Press, 2000), 149.

17. Turbayne, *Myth of Metaphor*, 22.

18. J. D. Barrow, *The World within the World* (Oxford, England: Oxford University Press, 1988), 70.

19. N. Cartwright, *The Dappled World: A Study of the Boundaries of Science* (Cambridge, England: Cambridge University Press, 1999); R. N. Giere, *Science without Laws* (Chicago: University of Chicago Press, 1999).

20. T. S. Kuhn, "Metaphor in Science," in *Metaphor and Thought*, 2d ed., ed. A. Ortony (Cambridge, England: Cambridge University Press, 1993), 539.

Chapter 3: The Theory of Conceptual Metaphor

1. D. E. Rummelhart, "Some Problems with the Notion of Literal Meanings," in *Metaphor and Thought*, 2d ed., ed. A. Ortony (Cambridge, England: Cambridge University Press, 1993), 71.

2. G. Lakoff, "The Contemporary Theory of Metaphor," in *Metaphor and Thought*, 2d ed., ed. A. Ortony (Cambridge, England: Cambridge University Press, 1993), 203.

3. G. Lakoff and M. Johnson, *Metaphors We Live By* (Chicago: University of Chicago Press, 1980).

4. For example, G. Lakoff and M. Johnson, *Philosophy in the Flesh: The Embodied Mind and Its Challenge to Western Thought* (New York: Basic Books, 1999).

5. Not all those interested in metaphor from the standpoint of linguistics or psychology are satisfied with the full range of claims made by Lakoff and Johnson. Here are a few references to alternative views: G. L. Murphy, "On Metaphoric Representation," *Cognition* 60 (1996): 173–204; G. L. Murphy, "Reasons to Doubt the Present Evidence for Metaphoric Representation," *Cognition* 62 (1997): 99–108; Gentner et al., "Metaphor Is Like Analogy."

6. Lakoff and Johnson, *Metaphors We Live By*, 98.

7. The terms "basic-level" and "superordinate" arise in theories of categorization. The basic-level category is that which optimally fits our perceptions of things as we distinguish them from one another and relate them. For example, in the general category "vehicle," "car," "truck," "train," and "airplane" are basic-level categories. The general category "vehicle" is the superordinate level. Below the basic-level category are more specific categories (e.g., below the basic level of "car" is the categorization in terms of manufacturer, color, number of doors, and so on). See Lakoff and Johnson, *Philosophy in the Flesh*, 26–30.

8. Lakoff and Johnson, *Metaphors We Live By*, 7, 8.

9. F. Waismann, "How I See Philosophy," in *Logical Positivism*, ed. A. J. Ayer (New York: Free Press, 1959), 347–48.

10. Lakoff, "Contemporary Theory of Metaphor," 216.

11. This interaction view of cognition owes much to Jean Piaget, Ernst Cassirer, and others.

12. M. Johnson, *The Body in the Mind: The Bodily Basis of Meaning, Imagination, and Reason* (Chicago: University of Chicago Press, 1987).

13. Lakoff and Johnson, *Metaphors We Live By*, 31.

14. Ibid., 57.

15. Ibid., chap. 15.

16. Johnson, *Body in the Mind*, 102.

17. K. Nauta and R. E. Miller, "Nonequilibrium Self-Assembly of Long Chains of Polar Molecules in Superfluid Helium," *Science* 283 (1999): 1895.

18. J.-W. Yoon, C.-S. Yoon, H.-W. Lim, Q. Q. Huang, Y. Kang, K. H. Pyun, K. Hirasawa, R. S. Sherwin, and H.-S. Jun, "Control of Autoimmune Diabetes in NOD Mice by GAD Expression or Suppression in β Cells," *Science* 284 (1999): 1185.

19. P. M. Platzman and M. I. Dykman, "Quantum Computing with Electrons Floating on Liquid Helium," *Science* 284 (1999): 1967.

20. J. Holash, P. C. Maisonpierre, D. Compton, P. Boland, C. R. Alexander, D. Zagzag, G. D. Yancopoulos, and S. J. Wiegand, "Vessel Cooption, Regression, and Growth in Tumors Mediated by Angiopoietins and VEGF," *Science* 284 (1999): 1995.

21. Lakoff and Johnson call these two basic metaphors for change the "location event-structure" metaphor and the "object event-structure" metaphor (*Philosophy in the Flesh*, chap. 11). I prefer to avoid these unwieldy names.

22. J. Pearl, *Causality: Models, Reasoning, and Inference* (Cambridge, England: Cambridge University Press, 2000).

23. A. P. Batista, C. A. Buneo, L. H. Snyder, and R. A. Anderson, "Reach Plans in Eye-Centered Coordinates," *Science* 285 (1999): 257.

24. K. Garber, "Taking Garbage In, Tossing Cancer Out?," *Science* 295 (2002): 612.

25. M. Chicurel, "Cell Migration on the Move," *Science* 295 (2002): 607.

26. S. Y. Hsu, K. Nakabayashi, S. Nishi, J. Kumagel, M. Kudo, O. D. Sherwood, and A. J. W. Hsueh, "Activation of Hormone Receptors by the Hormone Relaxin," *Science* 295 (2002): 671.

27. E. F. Kittay, *Metaphor: Its Cognitive Force and Linguistic Structure* (Oxford, England: Clarendon Press, 1987), 7.

28. It is worth noting that the number of people actually capable of measuring the speed of light is very small. The number capable of measuring it with the precision that leads to the accepted value is very, very small. Indeed, it is very likely that only a tiny fraction of scientists who use the number could even give a credible account of how the measurement is made or how the number can be known with the certainty with which it is reported. The point here is that scientists take a great deal of "fact" for granted. If scientific work is to progress, they have no choice in the matter. As a result, the objective nature of such canonical quantities as the speed of light is simply taken for granted. See Ian Hacking, *The Social Construction of What?* (Cambridge, Mass.: Harvard University Press, 1999), chap. 6.

Chapter 4: The Classical Atom

1. Lakoff and Johnson, *Philosophy in the Flesh,* 214.

2. Lucretius, *On Nature,* trans. R. M. Greer (New York: Bobbs-Merrill Co., 1965), 17.

3. Ibid., 45.

4. H. Butterfield, *The Origins of Modern Science,* rev. ed. (New York: Free Press, 1965).

5. S. Shapin, *The Scientific Revolution* (Chicago: University of Chicago Press, 1996), 96.

6. R. Boyle, *New Experiments Physico-Mechanical Touching the Spring of the Air, and Its Effects* (London, 1660), 12.

7. Insofar as our examination of the metaphors for atoms is concerned, it is not important that the gases studied historically, such as air, were composed of diatomic molecules (e.g., O_2) rather than single atoms.

8. I. Newton, *Optiks,* 2d ed. (London, 1719), 375–76.

9. Absolute zero does not correspond to an observable in the real world, even though today it is possible to approach absolute zero to within a few millionths of a degree.

10. Dalton was the son of a poor handloom weaver. His family were Quakers, and John studied at the Quaker school in his village. He left school at age eleven and opened his own school at age twelve. A few years later he taught at a cousin's academy and found opportunities to study with older masters. Somehow he managed in the course of all this to become a practicing research scientist. Eventually he found his way to Manchester, where he was able to learn of the work of Lavoisier and other researchers. One of his earlier proposals, in conflict with Lavoisier's beliefs, was that in a mixture of gases each gas behaves largely independently of the others. Thus, they remain chemically distinct.

11. S. Brush, "Introduction," in L. Boltzmann, *Lectures on Gas Theory,* trans. S. Brush (Berkeley: University of California Press, 1964), 1–17.

12. M. J. Nye, *Molecular Reality* (New York: American Elsevier, 1972).

13. L. Boltzmann, "On Certain Questions of the Theory of Gases," *Nature* 51 (1895): 413.

Chapter 5: The Modern Atom

1. In writing this chapter I have drawn extensively on P. F. Dahl's thorough historical treatment of the discovery of the electron as given in *The Flash of the Cathode Rays: A History of J. J. Thomson's Electron* (Bristol, England: Institute of Physics Publishing, 1997), as well as primary sources cited therein.

2. In his Nobel lecture in 1906, Thomson could not resist referring to the nationalistic nature of the competition: "Two views were prevalent; one, which was chiefly supported by the English physicists, was that the rays are negatively electrified bodies shot off from the cathode with great velocity; the other view, which was held by the great majority of German physicists, was that the rays are some kind of ethereal vibration or waves." J. J. Thomson, "Carriers of Negative Electricity," in *Nobel Lectures: Physics, 1901–1921* (Amsterdam: Elsevier, 1967), 146.

3. J. J. Thomson, "Cathode Rays," *Philosophical Magazine* 44 (1897): 295.

4. J. J. Thomson, "Carriers of Negative Electricity," 149.

5. To interpret the experimental results in terms of the internal constitution of the atom, Thomson had to resort to models for the interactions of the various kinds of rays with matter. Thus, the conclusions drawn, with implications for his metaphor for the constitution of atoms, were themselves based on a different metaphor accounting for the scattering observations. As scientists become increasingly dependent on complex instrumentation in making observations, the interpretations of the observations become correspondingly more dependent on the metaphorical models for how the instruments themselves work.

6. Dahl, *Flash of the Cathode Rays*, 322.

7. E. Rutherford, "The Scattering of Alpha and Beta Particles by Matter and the Structure of the Atom," *Philosophical Magazine* 21 (1911): 669–88.

8. Planck's theory of energy emission is a bit more complicated than this because the size of the "brick," or quantum, depends on the frequency of the radiation. Radiant energy is characterized by its frequency: High frequency means high energy, low frequency means low energy. Thus, radio broadcast frequencies are lower than infrared, which is lower in frequency than visible light, which is lower in frequency than ultraviolet, which is lower in frequency than X rays. The "bricks," or quanta, of radiant energy therefore would increase in size in this series, with the broadcast bricks the smallest and the X ray bricks the largest.

9. M. Planck, "The Genesis and Present State of Development of the Quantum Theory," in *Nobel Lectures: Physics, 1901–1921* (Amsterdam: Elsevier, 1967), 444.

10. T. S. Kuhn, *Black-Body Theory and the Quantum Discontinuity, 1894–1912* (Oxford, England: Oxford University Press, 1978).

11. The 1905 paper introducing the photon caused consternation among the greats of theoretical physics, including Max Planck. In 1913, Planck consigned a petition rec-

ommending Einstein for membership in the Prussian Academy of Sciences. However, the cosigners asked that Einstein's faulty paper on photon theory not be held against him!

12. N. Bohr, "On the Constitution of Atoms and Molecules," *Philosophical Magazine* 26 (1913): 1–25. Bohr wrote, "By the introduction of this quantity [Planck's constant] the question of the stability of the stable configuration of the electrons in the atoms is essentially changed as this constant is of such dimensions and magnitude that it, together with the mass and charge of the particles, can determine a length of the order of magnitude required" (2).

13. D. Wilson, *Rutherford: Simple Genius* (Cambridge, Mass.: MIT Press, 1983), 331.

14. W. Heisenberg, *Physics and Beyond: Encounters and Conversations* (New York: Harper and Row, 1971), 40.

15. G. P. Thomson was the son of J. J. Thomson. There is irony in J. J.'s son sharing in the Nobel prize (1937) for work demonstrating that the electron has the characteristics of a wave, precisely what his father had been at such pains to argue against forty years earlier. Of course, the wavelike character of the electron was argued on much different, and incorrect, grounds in J. J.'s day.

16. Quoted in B. Pullman, *The Atom in the History of Human Thought* (New York: Oxford University Press, 1998), 290.

17. Incidentally, it has been shown that the Schrödinger and Heisenberg matrix models are entirely equivalent in terms of the experimental results predicted.

18. Heisenberg, *Physics and Beyond.*

19. T. S. Kuhn, "Metaphor in Science," in *Metaphor and Thought*, 2d ed., ed. A. Ortony (Cambridge, England: Cambridge University Press, 1993), 538.

Chapter 6: Molecular Models in Chemistry and Biology

1. The story of the race to determine the correct structure of DNA, the molecule that encodes the hereditary information of living things, is recounted in James Watson's lively account, *The Double Helix* (New York: Atheneum, 1968), 197.

2. Pasteur was employed by the French wine industry to learn why wine often went sour while aging. He discovered that microorganisms produce the fermentation process that forms the wine. Heating the wine gently after fermentation, a process now called pasteurization, kills the microorganisms and prevents souring.

3. As it turned out, Pasteur had been lucky. The particular salt that he studied forms asymmetrical crystals only when crystallization occurs below a certain temperature, conditions that Pasteur had been fortunate to have happened upon.

4. When the Nobel prizes were first offered, in 1901, van't Hoff was the first recipient. His selection from a huge field of outstanding candidates is testimony to the extent to which he had enhanced the field of chemistry. However, he was not cited for his model of tetrahedral carbon but for work that in retrospect has been of less enduring importance.

5. V. A. Russell, C. C. Evans, W. Li, and M. D. Ward, "Nanoporous Molecular Sandwiches: Pillared Two-Dimensional Hydrogen-Bonded Networks with Adjustable Porosity," *Science* 276 (1997): 575–79.

6. T. M. Iverson, C. Luna-Chavez, G. Cecchini, and D. C. Rees, "Structure of the *Escherichia coli* Fumarate Reductase Respiratory Complex," *Science* 284 (1999): 1961–66.

7. The two major secondary structures of proteins, the α helix and the β sheet, were identified by Linus Pauling and Robert Corey in the early 1950s, on the basis of X ray diffraction studies of chains of amino acids.

8. Iverson et al., "Structure," 1966.

Chapter 7: Protein Folding

1. C. B. Anfinsen, "Principles That Govern the Folding of Protein Chains," *Science* 181 (1973): 223–30.

2. Ibid., 224.

3. For an interesting brief account of the discovery of the molecular nature of sickle cell anemia, see B. J. Strasser, "Sickle Cell Anemia, a Molecular Disease," *Science* 286 (1999): 1488–90.

4. F. M. Richards, "The Protein Folding Problem," *Scientific American,* Jan. 1991, 54–63.

5. D. Gentner and M. Jeziorski have laid out the details of Carnot's structure mapping analogy between water flow and heat flow in "The Shift from Metaphor to Analogy in Western Science," in *Metaphor and Thought,* 2d ed., ed. A. Ortony (Cambridge, England: Cambridge University Press, 1993), 447–80.

6. W. Kauzmann, "Some Factors in the Interpretation of Protein Denaturation," *Advances in Protein Chemistry* 14 (1959): 1–63.

7. B. Hayes, "Protototeins," *American Scientist* 86 (1998): 216–21.

8. K. A. Dill, S. Bromberg, K. Yue, K. M. Fiebig, D. P. Yee, P. D. Thomas, and H. S. Chan, "Principles of Protein Folding: A Perspective from Simple Exact Models," *Protein Science* 4 (1995): 561–602.

9. It is important to keep in mind that in this experiment any individual folded state is just as likely as any other individual unfolded state. But because there are so many more of the latter, the probability that the molecule will at any moment be in a folded form is very low.

10. C. Levinthal, "How to Fold Graciously" (with notes by A. Rawitch), in *Mössbauer Spectroscopy in Biological Systems,* ed. P. Debrunner, J. C. M. Tsibris, and E. Münck (Urbana: University of Illinois Department of Physics, 1969), 22–25.

11. G. Johnson, "Designing Life: Proteins 1, Computer 0," *New York Times,* 25 March 1997, B9.

12. J. N. Onuchic, Z. Luthey-Schulten, and P. G. Wolynes, "Theory of Protein Folding: The Energy Landscape Perspective," *Annual Reviews of Physical Chemistry* 48 (1997): 550.

13. J. N. Onuchic, N. D. Socci, Z. Luthey-Schulten, and P. G. Wolynes, "Protein Folding Funnels: The Nature of the Transition State Ensemble," *Folding and Design* 1 (1996): 441–50, available online at <http://www.BioMedNet.com/->; P. G. Leopold, M. Montal, and J. N. Onuchic, "Protein Folding Funnels: A Kinetic Approach to the Sequence-

Structure Relationship," *Proceedings of the National Academy of Sciences* 89 (1992): 8721–25.

Chapter 8: Cellular-Level Metaphors

1. R. Hooke, *Micrographia* (London, 1665).

2. RNA (ribonucleic acid) is a family of large molecules similar to DNA in composition. RNA performs many different functions in the life of the cell. One of its most important functions is to "read" the genetic information of DNA and direct the synthesis of proteins based on it.

3. "Frontiers in Cell Biology: Quality Control," *Science* 286 (1999): 1881–1905.

4. The solution did contain ionic substances that simulated the conditions of the cell. Such solutions are said to be at physiological ionic strength and acidity.

5. R. A. Laskey, B. M. Honda, A. D. Mills, and J. T. Finch, "Nucleosomes Are Assembled by an Acidic Protein Which Binds Histones and Transfers Them to DNA," *Nature* 275 (1978): 420.

6. J. Ellis, "Proteins as Molecular Chaperones," *Nature* 328 (1987): 378–79.

7. Anfinsen's original idea, that the order and identity of the amino acids in the protein chain determine the final native state structure, still stands. The chaperone protein facilitates folding, prevents misfolding, and prevents interactions with other folding proteins that would lead to undesirable aggregation, but in so doing it does not direct folding to a different structure.

8. S. Wickner, M. R. Maurizi, and S. Gottesman, "Post-Translational Quality Control: Folding, Refolding, and Degrading Proteins," *Science* 286 (1999): 1888–93.

9. G. Vogel, "Prusiner Recognized for Once-Heretical Prion Theory," *Science* 278 (1997): 214.

10. S. B. Prusiner, "Prion Diseases and the BSE Crisis," *Science* 278 (1997): 245–51; M. Balter, "Prions: A Lone Killer or a Vital Accomplice?," *Science* 286 (1999): 660–62.

11. S. J. Lippard, "Free Copper Ions in the Cell?," *Science* 284 (1999): 748–50; T. D. Rae, P. J. Schmidt, R. A. Pufahl, V. C. Culotta, and T. V. O'Halloran, "Undetectable Intracellular Free Copper: The Requirement of a Copper Chaperone for Superoxide Dismutase," *Science* 284 (1999): 805–8.

12. M. M. Waldrop, *Complexity: The Emerging Science at the Edge of Order and Chaos* (New York: Simon and Schuster, 1992); S. Kauffman, *At Home in the Universe: The Search for the Laws of Self-Organization and Complexity* (New York: Oxford University Press, 1995).

13. L. Helmuth, "Protein Clumps Hijack Cell's Clearance System," *Science* 292 (2001): 1467.

14. H. Li, S. K. Kolluri, J. Gu, M. I. Dawson, X. Cao, P. D. Hobbs, B. Lin, G.-Q. Chen, J.-S. Lu, F. Lin, Z. Xie, J. A. Fontana, J. C. Reed, and X.-K. Zhang, "Cytochrome *c* Release and Apoptosis Induced by Mitochondrial Targeting of Nuclear Orphan Receptor TR3," *Science* 289 (2000): 1159–64.

15. M. A. Takasu, M. B. Dalva, R. E. Zigmond, and M. E. Greenberg, "Modulation

of NMDA Receptor-Dependent Calcium Influx and Gene Expression through EphB Receptors," *Science* 295 (2002): 491.

Chapter 9: Global Warming

1. R. Bradley, "1000 Years of Climate Change," *Science* 288 (2000): 1353–55.

2. B. Wuethrich, "How Climate Change Alters Rhythms of the Wild," *Science* 287 (2000): 793–95; see also G.-R. Walther, E. Post, P. Convey, A. Menzel, C. Parmesan, T. J. C. Beebee, J.-M. Fromentin, O. Hoegh-Guldberg, and F. Bairlein, "Ecological Responses to Recent Climate Change," *Nature* 416 (2002): 389–95.

3. M. B. Dyurgenov and M. F. Meier, "Twentieth-Century Climate Change: Evidence from Small Glaciers," *Proceedings of the National Academy of Science* 97 (2000): 1406–11.

4. A. Rothrock, Y. Yu, and G. Maykut, "Thinning of the Arctic Sea-Ice Cover," *Geophysical Research Letters* 26 (1999): 3469–72.

5. R. A. Kerr, "Will the Arctic Ocean Lose All Its Ice?," *Science* 286 (1999): 1828; K. Y. Vinnikov, A. Robock, R. J. Stouffer, J. E. Walsh, C. L. Parkinson, D. J. Cavalieri, J. F. B. Mitchell, D. Garrett, and V. F. Zakharov, "Global Warming and Northern Hemisphere Sea Ice Extent," *Science* 286 (1999): 1934–37.

6. R. A. Kerr, "Globe's 'Missing Warming' Found in the Ocean," *Science* 287 (2000): 2126–27.

7. T. M. L. Wigley, "The Science of Climate Change: Global and U.S. Perspectives" (Pew Center on Global Climate Change, 2000), online at <http://www.pewclimate.org/>; D. Rind, "The Sun's Role in Climate Variations," *Science* 296 (2002): 673–77. For a contrarian view, see S. Baliunas, "Why So Hot? Don't Blame Man, Blame the Sun," *Wall Street Journal*, 5 Aug. 1999, A20.

8. A concentration of 375 ppm of carbon dioxide means that if a sample of air contained 1 million gas molecules, 375 of them would be carbon dioxide.

9. Actually, this model is only partly correct. The glass does act as described, but one of the major reasons that a greenhouse works as it does is simply that it confines the air and prevents ordinary exchange with air outside. This aspect of a greenhouse is not applicable to atmospheric greenhouse gases. Although it is by no means a perfect metaphor, the greenhouse has become an enduring one in discussions of global climate change.

10. This is not to suggest that these periods of glaciation were driven by variations in CO_2 levels. Other factors, such as the orbital variations mentioned earlier, are likely causes of episodes of glaciation. The role of CO_2 in glaciation episodes is not entirely clear. See P. U. Clark, R. B. Alley, and D. Pollard, "Northern Hemisphere Ice-Sheet Influences on Global Climate Change," *Science* 286 (1999): 1104–11.

11. A. Indermühle, E. Monnin, B. Stauffer, and T. F. Stocker, "Atmospheric CO_2 Concentration from 60 to 20 kyr BP from the Taylor Dome Ice Core, Antarctica," *Geophysical Research Letters* 27 (2000): 735–38.

12. Wigley, "Science of Climate Change," 23–25.

13. T. J. Crowley, "Causes of Climate Change over the Past 1000 Years," *Science* 289 (2000): 270–77.

14. *Forum on Global Change Modeling,* USGCRP Report 95-02, Washington, D.C., 1995, available from GCRIO User Services, 2250 Pierce Road, University Center, MI 48710; E. J. Barron, "Climate Models: How Reliable Are Their Predictions?," *Consequences* 1 (Autumn 1995), online at <http://www.gcrio.ciesin.org/Consequences/fall95/mod.html>.

15. M. E. Mann, R. S. Bradley, and M. K. Hughes, "Northern Hemisphere Temperatures during the Past Millennium: Inferences, Uncertainties, and Limitations," *Geophysical Research Letters* 26:6 (1999): 759–62.

16. P. A. Stott, S. F. B. Tett, G. S. Jones, M. R. Allen, J. F. B. Mitchell, and G. J. Jenkins, "External Control of Twentieth-Century Temperature by Natural and Anthropogenic Forcings," *Science* 290 (2000): 2133–37.

17. B. Hileman, "Consequences of Climate Change," *Chemical and Engineering News,* 27 Mar. 2000, 18; R. A. Kerr, "Dueling Models: Future U.S. Climate Uncertain," *Science* 288 (2000): 2113.

18. M. R. Allen, P. A. Stott, J. F. B. Mitchell, R. Schnur, and T. L. Delworth, "Quantifying the Uncertainty in Forecasts of Anthropogenic Climate Change," *Nature* 407 (2000): 617–20.

19. For a sampling of the contrarian views, see the Web sites of the Global Climate Coalition, <http://www.globalclimate.org>; Global Warming, <http://www.globalwarming.org>; and the George C. Marshall Institute, <http://www.marshall.org>.

20. Another example of such a problem is the question of how to manage radioactive wastes from nuclear power plants. A plan to place the wastes in underground storage under Yucca Mountain in Nevada has been subjected to innumerable studies. The issue is intensely political, with passionately held views on both sides. The scientific questions have to do with whether the wastes will remain confined to the storage space or leak out over time and possibly contaminate groundwater. Attempts to address this concern involve complex models that are subject to numerous uncertainties. There are no precise precedents to provide unequivocal guidance.

21. D. A. Schön, "Generative Metaphor: A Perspective on Problem-Setting in Social Policy," in *Metaphor and Thought,* 2d ed., ed. A. Ortony (Cambridge, England: Cambridge University Press, 1993), 137–63.

22. D. N. McCloskey, *If You're So Smart: The Narrative of Economic Expertise* (Chicago: University of Chicago Press, 1990), 6.

23. S. H. Schneider, *Laboratory Earth: The Planetary Gamble We Can't Afford to Lose* (New York: Basic Books, 1997), 74.

24. A. Robinson and Z. Robinson, "Science Has Spoken: Global Warming Is a Myth," *Wall Street Journal,* 4 Dec. 1997, A22.

Chapter 10: Science's Metaphorical Foundations

1. It seems very natural to speak the language of conceptual metaphor. Recently, when I was describing to a colleague the widespread use of a factory as metaphor for the cell, he looked at me rather quizzically and said, "But how else could you talk about it?" The implication seemed to be that such a view of the cell was inevitable; what was there re-

ally to talk about? I replied by asking him whether he could imagine how biologists raised in a culture in which they were insulated from any knowledge of factories would come to understand cell biology.

2. For a recent article on this subject, see B. Keysar, Y. Shen, S. Glucksberg, and W. S. Horton, "Conventional Language: How Metaphorical Is It?," *Journal of Memory and Language* 43 (2000): 576–93.

3. D. Gentner and D. R. Gentner, "Flowing Waters or Teeming Crowds: Mental Models of Electricity," in *Mental Models,* ed. D. Gentner and A. L. Stevens (Hillsdale, N.J.: Erlbaum, 1983), 99–129; D. Gentner and A. B. Markman, "Structure Mapping in Analogy and Similarity," *American Psychologist* 52 (1997): 45–56; Gentner et al., "Metaphor Is Like Analogy."

4. Lakoff and Johnson, *Metaphors We Live By;* Johnson, *Body in the Mind;* G. Lakoff, *Women, Fire, and Dangerous Things: What Categories Reveal about the Mind* (Chicago: University of Chicago Press, 1987); Lakoff and Johnson, *Philosophy in the Flesh.*

5. Lakoff, "Contemporary Theory of Metaphor."

6. R. Harré, *Varieties of Realism* (Oxford, England: Blackwell, 1986); H. Putnam, *The Many Faces of Realism* (LaSalle, Ill.: Open Court Publishing, 1987); J. Leplin, ed., *Scientific Realism* (Berkeley: University of California Press, 1984); P. Kitcher, *The Advancement of Science* (New York: Oxford University Press, 1993), chap. 5.

7. S. Glashow, "The Death of Science!?," in *The End of Science? Attack and Defense,* ed. Richard Q. Elvee (Lanham, Md.: University Press of America, 1992), 46–57.

8. L. Wittgenstein, *Philosophical Investigations* (New York: Macmillan, 1953).

9. F. J. Varela, E. Thompson, and E. Rosch, *The Embodied Mind: Cognitive Science and Human Experience* (Cambridge, Mass.: MIT Press, 1999).

10. A. Ortony, "Metaphor, Language and Thought," in *Metaphor and Thought,* ed. A. Ortony (Cambridge, England: Cambridge University Press, 1993), 1–16.

11. These remarks apply to technology as well as science.

12. J. A. Labinger, "The Science Wars and the Future of the American Academic Profession," *Daedalus* 126 (1997): 201–20.

13. A. Pickering, ed., *Science as Practice and Culture* (Chicago: University of Chicago Press, 1992); H. Collins and T. Pinch, *The Golem: What Everyone Should Know about Science* (Cambridge, England: Cambridge University Press, 1993); B. Latour, *Science in Action: How to Follow Scientists and Engineers through Society* (Cambridge, Mass.: Harvard University Press, 1987).

14. A. Pickering, *The Mangle of Practice: Time, Agency, and Science* (Chicago: University of Chicago Press, 1995); J. Golinski, *Making Natural Knowledge: Constructivism and the History of Science* (Cambridge, England: Cambridge University Press, 1998); R. Dunbar, *The Trouble with Science* (Cambridge, Mass.: Harvard University Press, 1995).

15. R. K. Merton, *The Sociology of Science,* ed. N. W. Storer (Chicago: University of Chicago Press, 1973), 446.

16. J. Ziman, *Real Science: What It Is and What It Means* (Cambridge, England: Cambridge University Press, 2000).

17. Lakoff and Johnson, *Philosophy in the Flesh*, 3.

18. G. Lakoff and R. E. Núñez, *Where Mathematics Comes From: How the Embodied Mind Brings Mathematics into Being* (New York: Basic Books, 2001).

19. G. R. Gribben, *The Search for Superstrings, Symmetry, and the Theory of Everything* (New York: Little, Brown, 1999).

20. R. B. Laughlin and D. Pines, "The Theory of Everything," *Proceedings of the National Academy of Science* 97 (2001): 28–31; R. B. Laughlin, D. Pines, J. Schmalian, B. P. Stojkovic, and P. Wolynes, "The Middle Way," *Proceedings of the National Academy of Science* 97 (2001): 32–37.

21. L. Daston, "Fear and Loathing of the Imagination in Science," *Daedalus* 127 (1998): 73–95.

INDEX

Theodore L. Brown is a professor emeritus of chemistry at the University of Illinois at Urbana-Champaign, where he also served as vice chancellor for research (1980–86), director of the Arnold and Mabel Beckman Institute for Advanced Research and Technology (1987–93), and interim vice-chancellor for academic affairs (1992–94).

The University of Illinois Press
is a founding member of the
Association of American University Presses.

Composed in 10/13 New Caledonia
with Helvetica Neue Extended display
by Jim Proefrock
at the University of Illinois Press
Designed by Dennis Roberts
Manufactured by Thomson-Shore, Inc.

University of Illinois Press
1325 South Oak Street
Champaign, IL 61820-6903
www.press.uillinois.edu